Selective Licensing

In recent years, the private rented sector has overtaken social housing to become the main housing provider with some of the worse housing conditions that are linked to preventable health inequalities. This book seeks to expand upon previous research in the area with a focus on selective licensing and enforcement, using a case study to illustrate changes in working practices that have been brought about through new powers being made available to local authorities to issue civil financial penalties upon criminal landlords.

The book examines the impact of this legislation on regulatory enforcement in the London Borough of Newham's property licensing scheme, delivered in a multi-agency partnership across its private rented sector, and the outcomes of combining the use of licensing and traditional housing inspections with use of civil penalties in alternative to prosecution to address some of the worse effects of poor housing.

The study also considers the limitations of employing informal actions to address such issues as well as identifying both the barriers to collaboration and the most effective strategies for service delivery where agencies – such as the police, border agency, council tax and local planning, irrespective of inter-agency competition – work together to achieve individual and shared objectives in evolving partnerships. The findings here will be of keen interest to environmental health professionals, academics, and indeed those operating in local authorities themselves.

Paul Oatt recently completed an MSc in Public Health Management at the London School of Hygiene and Tropical Medicine. He has worked in Environmental Health for over 17 years, mainly in housing carrying out HHSRS inspections and enforcement. He worked at Newham Council enforcing borough-wide licensing on their first five-year scheme. He now works as programme manager for Havering Council overseeing enforcement of their additional licensing scheme. Over the years, he has used many of the available enforcement tools to address serious housing hazards arising through poor management to ensure landlords either achieve compliance and reduce hazards to health or face legal consequences.

Routledge Focus on Environmental Health
Series Editor: Stephen Battersby, MBE PhD, FCIEH, FRSPH

Selective Licensing
The Basis for a Collaborative Approach to Addressing Health Inequalities

Paul Oatt

Routledge
Taylor & Francis Group

LONDON AND NEW YORK

First published 2020
by Routledge
4 Park Square, Milton Park, Abingdon, Oxon OX14 4RN
605 Third Avenue, New York, NY 10017

First issued in paperback 2023

Routledge is an imprint of the Taylor & Francis Group, an informa business

British Library Cataloguing-in-Publication Data
A catalogue record for this book is available from the British Library

Library of Congress Cataloging-in-Publication Data
Names: Oatt, Paul, author.
Title: Selective licensing: the basis for a collaborative approach to addressing health inequalities / Paul Oatt.
Description: First edition. | Abingdon, Oxon;
New York: Routledge, 2020. |
Series: Routledge focus on environmental health | Includes bibliographical references and index.
Identifiers: LCCN 2019044844 (print) | LCCN 2019044845 (ebook) |
ISBN 9780367429195 (hardback) | ISBN 9781003000136 (ebook)
Subjects: LCSH: Housing—England—London. |
Public health—England—London.
Classification: LCC HD7334.L6 O28 2020 (print) |
LCC HD7334.L6 (ebook) | DDC 362.1/042—dc23
LC record available at https://lccn.loc.gov/2019044844
LC ebook record available at https://lccn.loc.gov/2019044845

ISBN: 978-1-03-257092-1 (pbk)
ISBN: 978-0-367-42919-5 (hbk)
ISBN: 978-1-003-00013-6 (ebk)

DOI: 10.1201/9781003000136

Typeset in Times New Roman
by codeMantra

Publisher's Note
The publisher has gone to great lengths to ensure the quality of this reprint but points out that some imperfections in the original copies may be apparent.

Contents

Figures

Tables

Photographs (by kind permission of London Borough of Havering)

Series preface

This series, Routledge Focus on Environmental Health, is now maturing and is no longer just an initiative. The aim remains the same; to explore environmental health topics traditional or new and sometimes contentious issues, in more detail than might be found in the usual environmental health texts.

We want to encourage readers and practitioners, particularly those who might not have had work published previously to submit proposals as we hope to be responsive to the needs of environmental and public health practitioners. I am very keen that this is seen as an opportunity for first-time authors and as such would urge students (whether at first or second degree level) to consider this an avenue for publishing findings from research. Why for example should the hard work that has gone into a dissertation lie in an unread book on a library shelf?

Our hope remains that this is a dynamic series, providing a forum for new ideas and debate on current environmental health topics. So if you have any ideas for monographs in the series, please do not be afraid to submit them to me as series editor via the e-mail address below.

I have always encouraged new authors and for environmental health practitioners on the front line to "get published" writing from their experiences of trying to protect public health. Setting down in writing some analysis of what worked in practice, what was successful, what wasn't and why, can provide useful insights for others working in the field. It is not just an exercise in gathering CPD hours, but can provide a useful method of reflection and aid career development, something that anyone who considers themselves a professional should do. Furthermore, all too often the work of EHPs goes unrecorded and unremarked and with the demise of the *Journal of Environmental Health* I am pleased to be working with Routledge to provide this opportunity as another route for practitioners to change this.

It is not intended that this series take a wholly "technical" approach but provide an opportunity to consider areas of practice in a different way, for example, looking at the social and political aspects of environmental health in addition to a more discursive approach on specialist areas.

We recognise that "environmental health" can be taken to mean different things in different countries around the world. I know that Clay's *Handbook on Environmental Health* has chapters that might not be relevant to some practitioners in different countries; nevertheless, EHPs are a key part of the public health workforce wherever they practise. So, this series will enable a wider range of practitioners and others with a professional interest to access information and also to write about issues relevant to them. The format means a relatively short production time so contents will be more immediate than in a standard textbook or reference work.

Forthcoming monographs are likely to cover such areas as Environmental Health in South Africa, and Air Pollution and Health. That does not mean we have no need of further suggestions, quite the contrary, so I hope readers with ideas for a monograph will get in touch via Ed.Needle@tandf.co.uk or Patrick.Hetherington@tandf.co.uk.

Stephen Battersby MBE PhD, FCEIH, FRSPH
Series Editor

Acknowledgements

I would like to thank everyone who has supported me in the research and the making of this book and to the many hardworking environmental health practitioners (EHPs) I have had the pleasure of working with and learning from over the past 17 years of my journey in this profession. This also includes the many talented lecturers at Middlesex University who trained me in the necessary skills to analyse and investigate environmental hazards across the five disciplines, the tools of the EHPs trade.

As an EHP one is expected to carry out an annual quota of continued professional development (CPD), encouraging a culture of lifelong learning. In this spirit, I decided to continue my study by taking a master's in Public Health at the London School of Hygiene and Tropical Medicine, and I owe an immense debt of gratitude to the many public health professionals I was able to learn from, lecturers and students alike.

I would also like to thank the London Borough of Newham and the London Borough of Havering for their support of my career development and for allowing use of photographs in this publication.

Particular mention must go to Russell Moffatt, Dr Harry Rutter, Dr Stephen Battersby, Dr Jill Stewart, Dr Jennifer Gosling, Professor Valerie Iles, Dr Alan Page, Charles Seechurn Ruth Plume, Louise Watkinson, Sasha Taylor and Laura Burkin.

Thanks also to Ed Needle and Patrick Hetherington at Routledge, Taylor and Francis, for their invaluable guidance through the publication process. The policy analysis was taken from my dissertation at the London School of Hygiene and Tropical Medicine (Oatt, 2017). What are the ways in which collaborative working in licensing of private rented housing in the London Borough of Newham could be most effectively used to tackle the worst effects of poor housing? Some of the literature review is also adapted from this source; however, at the time of writing it up during 2017, housing policy and legislation was going through many changes and the section has now been updated to reflect this.

Lastly, I would like to thank my wife Jenny and my daughter Phoenix for all their support and without whom this would not have been possible.

Abbreviations

ASB	Antisocial Behaviour
CIEH	Chartered Institute of Environmental Health
FPN	Civil Financial Penalty Notice
EQIA	Equality Impact Assessment
EHP	Environmental Health Practitioners
FTT	First Tier Tribunal Residential Property Chamber
HHSRS	Health and Housing Safety Rating System
HMO	House in Multiple Occupation
IMD	Index of Multiple Deprivation
JSNA	Joint Strategic Needs Assessments
LAA	Local Area Agreements
LSPs	Local Strategic Partnerships
LSOA	Lower Super Output Area

1 Introduction

Like many environmental health practitioners (EHPs), I studied environmental health across the five disciplines of health and safety, food, pollution, contaminated land and housing choosing to specialise in a housing career because I feel passionately that to get strong working housing enforcement policies right would allow us to really turn the corner on public health. It is surprising to think that we can go out and eat and be confident that a food business with a high star rating is not going to poison us and that we can go to work and be reasonably confident that if we follow health and safety procedures we are not going to get injured. But coming home afterwards to a rented property in large parts of our country could mean a higher exposure to hazards in the very place where we are meant to feel most comfortable and secure.

Since the enablement of the enforcement powers under the Housing Act 2004, there are still many local authorities seeking to address category 1 hazards through informal action. EHPs have status and autonomy to 'problem solve' their cases, yet prevalence of informal action is inconsistent with the law and undermining to those officers who seek to enforce under a consistent legally based approach. The first section of this book examines reasons why informal action is prevalent, suggesting it is rooted in poor management and poor understanding of the development and training needs of officers as well as a lack of understanding or regard to the aims and objectives of housing enforcement. Ultimately it fails to change criminal landlord behaviour.

The enforcement under part 1 is usually carried out reactively by way of response to occupier's complaints, whilst licensing under parts 2 and 3 provides a platform for local authorities to use licensing conditions to improve landlord's own management practices towards their own properties and improve standards through coordinated multi-agency enforcement. But this must all be underpinned by strong management of enforcement staff.

The case study illustrates a combined approach to using the act appropriately to address licensing and regulation breaches as well as category 1 hazards. The third section examines these different aspects in further detail using the legal analysis made in the recent *Brown v. Hyndburn* case where an appeal was made by a landlord who argued the council were using licensing conditions to require improvements, which is a function of part 1 of the act. Use of part 1 powers alone as a reactive enforcement model is ineffective and resource-intensive, whereas a combined approach with licensing conditions allows for a more effective model of enforcement. But in order to devise a robust enforcement policy underpinned by licensing, it is crucial to understand the aims of the different parts of the act, why they are separated into parts 1, 2 and 3, and how to make effective use of them.

The fourth section details a policy analysis of Newham's first five-year borough-wide licensing scheme which was undertaken at the midpoint of the project using Morestin's 'Framework for Analyzing Public Policies', taking account of policy context, justifications and supporting data, using a structured analysis approach developed from a public health perspective.

Finally, a literature review was carried out, and the findings showed that there is not a large amount of literature on licensing and housing enforcement interventions, illustrating the need for more literature on the subject, and this book I hope is a step forward in that direction. The review examines partnership and multi-agency working for which it was necessary to look at other types of public health and housing interventions involving partnerships with other agencies. The lack of available published material from local authorities is largely because enforcement functions have always been seen as a statutory duty, and it is only in recent years where local authorities are forced to make tough decisions over services that the need for evidence-based justifications has come to the fore.

Selective licensing is a tool that can be used as part of a wider evidence-based public health and housing policy. The evidence gathered in joint strategic needs assessments, housing stock condition surveys and other types of evidential analysis has wider uses, and a sharing of resources across services to develop a combined approach to tackling homelessness, antisocial behaviour, poor housing management, poor housing conditions, overcrowding, deprivation and crime starts with the development of shared goals across services underpinned by strong partnerships and support of good staff willing and able to tackle these issues head on.

2 Don't cross that line!

Any discussion about housing enforcement inevitably centres upon the application of enforcement legislation under the Housing Act 2004 (hereinafter referred to as 'the act'). This statutory instrument had enabled environmental health practitioners (EHPs) to address poor housing conditions '*Arising from a deficiency in the dwelling...as a result of the construction... [or] absence of maintenance and repair,*' where exposure to such deficiencies presents '*a risk of harm to the health and safety of ... occupiers*' (p. 3).[1] The most serious classification of this hazard type is a category 1 hazard.

The act states that a local authority '*must*' take '*appropriate enforcement action*'(p. 5),[1] where a category 1 hazard exists, to serve statutory notices in order to eradicate hazards to health. In 2016, *Environmental Health News*[2] reported that local councils rarely use housing powers and are more likely to '*take informal action*' such as the '*sending of a warning letter,*' in place of exercising their '*statutory duty.*'

Around the same time, Campbell Robb,[3] then Chief Executive of housing and homelessness charity, Shelter, stated in an interview with the Guardian that '*conditions in many rented homes are as bad as we've seen in decades.*' The most common types of hazards in homes according to Shelter[4] are damp and mould, pests or cold-related issues arising from ineffective heating or poor insulation, along with safety hazards arising from electrical faults, inadequate fire safety or lack of carbon monoxide detection.

The basis of the *Environmental Health News* article originated from the findings of Dr Stephen Battersby's survey of local housing authorities.[5] The purpose of the survey was '*to assess the extent of any change in activity ... particularly as the private rented sector has grown at a time when financial cuts are affecting local authorities.*' The survey showed that between 2012 and 2014 private rented sector inspection rates decreased amidst an increasing prevalence of the category 1 hazards of damp and mould, crowding and space, falling hazards and fire safety, thus bearing out Shelters' claims. According to the survey findings, informal action rates increased from 13,754 in 2011 to 15,964 in 2015 (pp. 3–6).[5]

This begs the question, why are EHPs not using the tools available to them, to their full capacity? EHPs have professional status and recognised skills. The pathway to becoming an EHP involves studying the subject of environmental health at degree level, and once you graduate you must still pass further exams under the Chartered Institute of Environmental Health (CIEH) to earn the protected title of EHP and thus demonstrate competency in the application of your acquired skills for the protection of public health in whichever of the five environmental disciplines you work in – food, health and safety, pollution, contaminated land or housing.

At the university, I learned the necessary skills needed to analyse and investigate environmental hazards across the five disciplines, the tools of the EHP's trade. As an EHP one is expected to carry out an annual quota of continued professional development, encouraging a culture of lifelong learning and application of skills based on current knowledge.

Iles argues that society *'expects professionals to use their status…in the service of society.'*[6] But society having conferred status has difficulty challenging professionals *'…when they are not using their status in the interests of society.'*[6] This is partly because the knowledge can seem esoteric to the layperson, and it is hard for society to know whether or not the decisions made by professionals are correct, in much the same way that when most of us take our car for a service we are reliant on the mechanic to explain what work needs to be carried out, or when a doctor gives a diagnosis and a prescribes a form of treatment. I am sure that most of the time the information is correct to the best of current available knowledge, but we as laypersons place reliance on the expert to use their esoteric knowledge to act in our best interests.

Informal action is not prescribed as appropriate enforcement under the act (p. 5).[1] The act's intention is to address category 1 hazards via enforcement. There is a right of appeal via the First Tier Tribunal Residential Property Chamber (FTT), and prosecutions are heard at the magistrates court. One disadvantage during appeals is that the notice becomes suspended pending the outcome, so occupiers continue to be exposed to hazardous conditions until the appeal is heard. The prospect of representing the local authority in these matters is daunting and incentivises preference for informal action rather than entering into litigation with landlords. Conversely, issuing informal demands for works carries no right of appeal, and has no basis in legality.

In 2018, the *Guardian* reported on the findings of a freedom-of-information request to survey all 349 local authorities in England and Wales about their housing prosecutions, only to find that 53 councils had not prosecuted a single landlord in three years.[7]

I was taught how to survey a property and risk assess, using the Health and Housing Safety Rating System (HHSRS) – a risk-based evaluation tool used to determine the category of hazard that occupiers were exposed to through housing defects. In making these decisions, government guidance states that local authorities should give clear advice to a landlord on what is required of them, providing an opportunity to discuss the circumstances of a case before formal action is taken.[8]

This is consistent with government's guidance on the principles of good enforcement and policy procedures, to ensure that they were advised in writing in an easy-to-understand manner explaining the need for work to be carried out and over what time scale, with an opportunity to discuss the case, narrow down the issues and clarify them before taking formal action.[9]

Battersby's survey found that local authorities are breaching their statutory duty through the use of informal means, such as writing a letter to a landlord instead of serving a notice, rather than using it as a prerequisite to taking formal action (p. 11).[5]

This always puts me in mind of the old Warner Brothers cartoons with the '*don't cross this line*' routine, which goes something like this – two cartoon characters face off against each other. One draws a line in the sand between them and says, '*don't you dare cross it or you'll be sorry!*' This is like the council informally telling a landlord to carry out works to repair their property, because if they don't then the council will give '*consideration to carrying out enforcement action.*' In the cartoon, the character who has been challenged inevitably crosses the line, and immediately another one is drawn, '*right I'm warning you, don't cross this line or else.*'

To me this is akin to the situation whereupon the landlord has ignored the letter and absent response, the council re-inspects the property and finds it to be still in the same sorry state –full of damp and mould, inadequate heating, lack of fire safety and so on. But instead of serving a notice, the officer opts to write another warning letter, strongly advising the landlord that the works must be carried out over a specified time period or else the council will give '*serious consideration to carrying out enforcement action,*' and this dance can go on and on exactly like it does in the cartoons, drawing line after line in the sand, without a proper resolution.

Informal action, in my experience, does not resolve housing matters satisfactorily. It has its place only within a finite consultation period where the EHP inspects under HHSRS and makes the landlord aware of the findings within a minimum two-week consultation period before serving the appropriate notice. This is the practice I follow because I have the status and the autonomy to do so. However, this approach may not necessarily be consistent across a whole team, and such enforcement efforts are undermined by other officers who rely only upon informal action, drawing lines in the sand, prolonging the process indefinitely without exercising the appropriate formal action. Some landlords have portfolios of properties across boroughs, and often they are simultaneously dealing with different officers in relation to problems found in their properties. It is understandable that they would find the inconsistency of approaches used by different officers confusing.

In my view, the root of the problem lies in poor management. Managers are not properly challenging the decisions of their staff. The relationship between managers and EHPs and the relationship between EHPs and landlords and tenants are key factors in the management of housing enforcement. Iles gives three basic rules for managing people that apply equally. First, there should be agreement between both parties. Managers and EHPs can have broad or specific conversations on what the service and individual officers are expected to achieve such as agreement over the need to carry out statutory duty to protect public health (p. 471).[10]

Managers can tailor such conversations to the personalities of staff members, either reaching agreement on specific service vis-à-vis measurable outcomes or those that are measurable through individual performance. Environmental health is a holistic discipline. There is scope to converse about the bigger picture,[10] that is, how enforcing repairs in the private rented sector benefits vulnerable occupiers and society as a whole, helping to reduce the costs to the NHS for the treatment of preventable injury or illness, a crucial line of argument when applying for funding (p. 7).[11]

The second rule for managing people is to '*ensure you are both confident they have the skills and resources.*' EHPs work autonomously off site. For office-based managers overseeing staff working externally, it would be far more supportive to periodically accompany EHPs, and observe them in action, because whilst analysis and risk assessment are taught academically, they are skills that need to develop with practice (p. 471).[10]

First-hand observation leads naturally into rule number three – providing ongoing feedback.[10] EHPs performance is best measured by observing how the officer manages site visits, carries out surveys and risk assessments, and develops their communication skills, engaging with landlords and tenants.

Councils often monitor progress using reports of numbers of tenant complaints (inputs) and responses (outputs). This is simply a measure of demand and supply and is often erroneously applied to measuring individual performance. It contributes little to discussion on development of the officer's skills or their reflection on the handling of significant or problematic cases. Iles emphasises the need for balance to ensure there is more positive feedback than negative, so that *'the other person leaves the conversation feeling that they want to make changes, and are confident in their ability to do so.'*[10]

Between the landlords who respond positively to informal requests for action and those who ignore the council, there are those who will look at the list of works that the council has prescribed informally and argue that they need vacant possession of the property first, after which they promise to carry out the work when the tenancy ends. Some of them keep to their word, but a great deal of them don't and rely on the fact that your service is overstretched hoping that their case will simply be forgotten. When your backs turned, they simply cover up the problems, that is, paint over the mould and re-let to another unsuspecting tenant. Is this approach sustainable? Does it change behaviour? The honest answer is no.

Local authorities that regulate the private rented sector without selective or additional licensing schemes only have the powers under part 1 of the act available to them. Further chapters will explore the different intentions behind parts 1, 2 and 3 of the act, as well as illustrate through means of a case study a combined use of these powers.

In 2010, Sir Michael Marmot had published a review into health inequalities in England proposing an evidence-based strategy to address the social determinants of health, that is, the conditions in which people are born, grow, live, work and age which can lead to health inequalities.[12]

Marmot's review called for more housing investment 'across the social gradient,' citing overcrowding, and 'poor physical condition' as being amongst the causal factors of preventable health inequalities. In 2010, overcrowding in Newham was found to be prevalent in 23.3% of the private rented sector, particularly within Newham's Manor Park area,[13] which also has a high proportion of HMOs and houses subdivided into smaller flats.[14]

Marmot argued for collaborative partnerships to be channelled through local authorities, health services, private sector and other groups to address these issues.[12]

To me this method of working seemed consistent with licensing regulation exemplified by Newham's approach, underpinned by the belief that sustainable improvements in housing conditions can be delivered by ensuring that landlords are fully monitored through the objectives of their licensing scheme set out as follows:

- '*Anti-social behaviour (ASB) is dealt with effectively*
- *Tenants' health, safety and welfare are safeguarded*
- *Landlords are "fit and proper persons" or employ agents who are*
- *Adequate property and tenancy management arrangements are in place*
- *Accommodation is suitable for the number of occupiers*
- *All landlords and managing agents operate at the same minimum level of professional standards.*'[15]

Between February 2013 and November 2016, 368 operations were carried out, using powers of entry into rented properties under section 239(7) of the act, equating to approximately two operations per week.[16]

I was part of a team that carried out 908 landlord prosecutions during that period, tasked with investigating private rented properties to ascertain whether offences have been committed by inspecting to ensure that licence conditions and housing regulations are being upheld and that properties are kept in a safe condition absent of hazards to health for occupiers.[16]

Evidence gathered through these operations was intended to maximise efficient use of officer time, by avoiding duplication between collaborating agencies.[17] When we drew a line in the sand you crossed it at your peril.

References

1. legislation.gov.uk. *Housing Act 2004* [online]. Available from: www.legislation.gov.uk/ukpga/2004/34/contents [Accessed 27th February 2016.]
2. Wall T. *Housing Enforcement Powers Rarely Used*. EHN [online]. (11th January 2016). Available from: www.ehn-online.com/news/article.aspx?id=15087 [Accessed 27th February 2016.]
3. Doward J. *Housing crisis reaches 1960s levels as tenants battle to cope, says shelter*. The Guardian [online]. (7th January 2016). Available from: www.theguardian.com/society/2016/jan/09/housing-crisis-tenants-shelter-private-rent [Accessed 27th February 2016.]
4. shelter.org.uk. *Health and safety problems in rented housing* [online]. (2015). Available from: http://england.shelter.org.uk/get_advice/repairs_and_bad_conditions/health_and_safety/health_and_safety_assessments_of_rented_homes [Accessed 27th February 2016.]
5. Battersby S. *The challenge of tackling unsafe and unhealthy housing* [online]. (December 2015). Available from: http://sabattersby.co.uk/documents/KBReport2.pdf [Accessed 27th February 2016.]

6. Iles V. *Working with others.* [Lecture / Presentation] Organisational Management. London School of Hygiene and Tropical Medicine, 23rd February 2016.

7. Wall T. *53 councils have not prosecuted a single landlord in three years* The Guardian [online]. (24th October 2018). Available from: www.theguardian.com/business/2018/oct/24/53-councils-have-not-prosecuted-a-single-landlord-in-three-years [Accessed 22nd August 2019.]

8. Assets.publishing.service.gov.uk. *Housing health and safety rating system guidance for landlords and property related professionals* [online]. (2006). Available from: https://assets.publishing.service.gov.uk/government/uploads/system/uploads/attachment_data/file/9425/150940.pdf [Accessed 3 August 2019.]

9. Publications.parliament.uk. *House of lords – delegated powers and de-regulation - twenty-eighth report* [online]. (2019). Available from: https://publications.parliament.uk/pa/ld199899/ldselect/lddereg/111/11107.htm [Accessed 3 August 2019.]

10. Iles V. Managing People and Teams. In Walshe K, Smith J, (eds.). *Healthcare Management.* Second Edition. Maidenhead: Open University Press, 2011, pp. 470–487.

11. Nicol S, Roys M, Garrett H. The Cost of Poor Housing to the NHS. BRE [online]. (2015). Available from: www.bre.co.uk/filelibrary/pdf/87741-Cost-of-Poor-Housing-Briefing-Paper-v3.pdf [Accessed 27th February 2016.]

12. Marmot M. *Fair society, healthy lives: the marmot review: strategic review of health inequalities in England post 2010,* 2010.

13. London Borough of Newham and NHS Newham. Joint strategic needs assessment 2010. Chapter 11 – Housing [online]. (11th January 2011). Available from: www.Newham.info/Custom/JSNA/Chapter11Housing-Newham.pdf [Accessed 21st December 2016]

14. London Borough of Newham. *Newham 2027 Newham's Local Plan the Core Strategy* [online]. (April 2013). Available from: www.Newham.gov.uk/Documents/Environment%20and%20planning/CoreStrategy2004-13.pdf [Accessed 21st December 2016.]

15. London Borough of Newham. Private rented property licensing guide for landlords and managing agents [online]. (June 2012). Available from: www.Newham.gov.uk/Documents/Housing/PropertyLicensingGuide-LandlordsAndManagingAgents.pdf [Accessed 21st December 2016.]

16. London Borough of Newham. Rented property licensing, proposal report for consultation [online]. (October 2016). Available from: www.newham.gov.uk/Documents/Housing/RentedPropertyLicensingProposalConsultation.pdf [Accessed 26th August 2019.]

17. Mishkin P, Moffatt R. A review of multi-agency enforcement and discretionary property licensing to tackle Newham's private rented sector. In Stewart J (ed.) *Effective Strategies and Interventions: Environmental Health and the Private Housing Sector.* London, England: Chartered Institute of Environmental Health, 2013, pp. 12–14.

3   The case study

Introduction

There are two central components to this intervention: (a) a multi-agency inspection under licensing enforcement, resulting in the issuing of financial penalties, and (b) an HHSRS (Health and Housing Safety Rating System) inspection that resulted in the service of a prohibition order.[1]

An occupier complained about not having gas or hot water between 9 a.m. and 5 p.m., saying that she 'rents a room and has a 3-month-old baby.' I was concerned about the occupancy level of the property and its conditions based on the occupier's complaint. The lack of availability of daytime heating and hot water was particularly worrying, suggesting these amenities were being withheld, and worse, the thought of a three-month-old baby living in those conditions.[1]

Borough-wide licensing schemes make it relatively easy to access records and find out the licence holder's details. I noted that the licence was held by the freeholder who was granted a selective licence for a single household based on the room sizes and facilities. The licence holder had also appointed an agency to manage the property. The only other record of note was a visit by planning in 2012, when at that time there were suspicions that an outhouse was being used as living space. The visit confirmed the house was a House in Multiple Occupation (HMO) occupied by nine persons, but the outhouse was used for storage.[1]

The issues raised suggested the property was poorly managed. In my experience, these circumstances, together with high rents and the demand for rooms in London, can all increase the risk of a large property of this type becoming an HMO through subletting. For these reasons, I wanted to verify that the outhouse was not being used for living purposes, and believed it necessary to inspect the property.[1]

Licence condition audit

Newham has adopted a policy of issuing checks to audit compliance with licence conditions, referred to as 'condition audits,' involving writing to licence holders to request various documents be produced as proof that the property is suitably maintained and managed, ensuring that licence holders are complying with the conditions of the licence and to drive up management standards in the sector. Licence holder's duty to respond to requests for documents within 28 days is set in the conditions (pp. 32–36).[2]

The outcome of the audit is used to gauge the level of licence holder's engagement with the property. Where there is no response to the condition audit, or the response is highly unsatisfactory, this is used as an indicator that the licence holder is not exercising proper management of the property and further investigation is carried out. A condition audit letter was sent to the licence holder, but no response was received during that time.

Multiagency inspection

Section 239(7) of the act[3] articulates the powers of entry for local authorities to inspect and determine if any licensing or HMO management regulation offences have been committed. Such inspections can be carried out at any reasonable time and without having to give any prior notice, allowing environmental health practitioners (EHPs) the opportunity to see the property in its true state. Evidence gathered in this way maximises efficient use of officer time, avoiding duplication between collaborating agencies,[4] to help bring about improvement in existing private rented sector housing quality stock across tenures and to reduce the adverse impact on peoples' health and personal development.

The property was a two-storey mid-terrace Victorian house in the Forest Gate area, with two-storey back addition and a ground floor rear extension (as viewed from street facing the property; see Figure 3.1).

The ground floor front room was in use as a living room, and the ground floor rear left room was being used as a bedroom. Within the back addition and extended area were a kitchen, small shower room and a separate bathroom divided by a lobby. On the first floor were three bedrooms, one in the first-floor back addition, one at the first-floor rear left side, and a first-floor front main bedroom.

Through a combination of interviews, collection of witness statements and verification of identification I established that the property

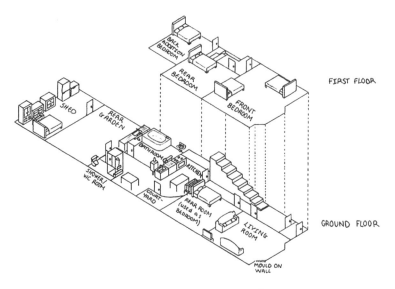

Figure 3.1 Exploded isometric property floorplan.[5]

is occupied by a head tenant who I will call Yasmin, and her four children who sleep in the first-floor front and rear left-side rooms. The ground floor rear left room and first-floor back addition were occupied by two single males who claimed to be just 'guests.'

'Yasmin' confirmed her rent is mostly paid by housing benefit which she tops up weekly and identified the landlord as the managing agent whom I found named on the licence; let's call them 'H Property Management.' I was satisfied from rent receipts showing this information that the agent details matched those that we have on record for the licence, as the landlord's appointed agent. Yasmin told me that she recently saw the landlord because they came to speak to the 'other family' living outside; at this point, she gestured towards the outhouse in the garden.

I went to this outhouse and interviewed and took a witness statement from 'Maya' and established that she was our complainant and had been living at the property for four or five months with her husband and baby.[1]

At this point, it was apparent that the property was being used as an HMO. The property was licensed under section 95 of the act,[3] as property requiring a selective licence for one household. The property consisted of more than one household, thus constituting a breach of licence conditions because it was only licensed for one household. The licence conditions in relation to permitted occupancy are that a new resident must not be permitted to occupy the house or any part of the house if that occupation exceeds the maximum permitted number of households (p. 31).[2]

Because of this breach, the property is not licensed as an HMO and should have been so licensed. This is an offence under section 72(1) of the act for failure to licence an additional HMO.[3]

The backdoor of the main house leading to the garden had a fixed glaze window that had been smashed in from outside, leaving jagged glass shards in the window and a hole wide enough for a person outside to reach in and unbolt the door. Because the property has the wrong licence and is an HMO, and because of the state of this door being in a common part of the property, the failure to carry out the repair constitutes an offence under the management regulations of HMO's section 234(3) of the act which impose duties on the person managing the house for their failure in complying. The regulations themselves are set out in The Management of Houses in Multiple Occupation (England) Regulations 2006, and the offence is under regulation 7(2)(d), which stipulates '*all windows and other means of ventilation within the common parts are kept in good repair.*'[6]

According to Maya, the broken window occurred because Yasmin's family locked the backdoor during the day, leaving Maya and baby trapped in the garden with no access to the kitchen or bathroom, or means of exit from the property. It is my belief that Maya experienced harassment from Yasmin and family who wanted them to leave.

It was clear to me that whilst these offences had already been evidenced during the inspection, a further inspection under HHSRS would need to be carried out in order to deal with the shed.

HHSRS inspection

One week after the multi-agency inspection, I was back at the property to carry out the HHSRS inspection.[1]

The shed was a detached single-storey structure built approximately within 0.5 m to the boundaries of three adjoining properties (adjacent left and right properties and rear property) with a floor area of 13.3 m². The structure is that of a breezeblock-built outhouse with a suspended timber floor and flat felted roof. The walls are approximately 100-mm thick being a single skin of breezeblock with a surface skim plaster finish on the interior. The roof is flat with no way of telling if the void in the roof space is filled with insulation. The dwelling itself consists of a small bedroom which also functions as a living area. The only form of heating is a 1 kW freestanding electric convector heater. There are gaps in the flooring allowing heat escape and draughts. Water marks at ground level on the breezeblock told me there was rising damp owing to the porosity of the breezeblock, further compromising what little thermal value the construct offered, and there was no drainage outlet in the garden and the threshold allowed water to enter.[1]

Inside the shed, the single plug in heater lacked sufficient capacity to achieve and maintain an adequate indoor temperature throughout the dwelling during cold weather; occupiers sleeping in these conditions are at a substantially increased risk of exposure to cold air streams with potential to affect the respiratory tract risking temporary slowing of the heart and increasing cardiovascular strain. When the whole body is cooled in this way, blood pressure increases. The effect of cool air on the bronchial lining and immune system can reduce resistance to infection (pp. 55–58).[7]

I was also concerned about fire safety because the family would have no early warning of fire and would face a complicated escape route through the main house to reach safety. There was haphazard electrical wiring in the shed that also heightened the risk. In terms of the escape route, the garden is mid-terrace and has no side passageway; it is flanked by high walls made out of the same breezeblock construction, and it would be difficult for a vulnerable person to seek escape by scaling the garden walls.[1]

Category 1 and 2 hazards are then defined in the act as 'hazards of a prescribed description' falling within a 'prescribed band' or numerical score which is arrived at by a prescribed method of calculating the seriousness of the hazard and accounts for the likelihood and severity of harm. A hazard refers to risk or harm to the occupiers (actual or potential occupiers) health or safety arising from a deficiency in the building or dwelling or absence of maintenance or repair or otherwise.[3]

I scored the hazards later back at the office. Excess cold and fire safety were found to be the main category 1 hazards whilst damp and mould and inadequate provision of drainage were high category 2 hazards. Clearly, a strongly worded informal letter and a line in the sand were not going to cut it. If the family moved out, that would not be sufficient to close the case; another family could well be moving in next week if I didn't act.[1]

The Police and Criminal Evidence Act (PACE) 1984 is used by anyone who investigates and prosecutes criminals, not just the police, including EHPs and other types of investigatory council officers. It is used for the questioning of persons of whom there are grounds to suspect of an offence. Before such a person is questioned regarding their involvement in the offence they must be cautioned.[8]

The licence holder whom I shall call Mr K was present on the day of the inspection. I suspected he was guilty of breaching his licence conditions under section 95(2) of the act failing to licence an HMO under section 72(1) and the management regulations of HMO's section 234(3).[3]

Under caution Mr K confirmed that H Property Management were given the property to manage a year ago; I was satisfied that this confirmed the managing arrangements as set out in the licence. Mr K told me he still visits the property, and last visited two or three days ago.[1]

Mr K stated that two or three months back he had asked H Property Management if everything was okay with the property and was told that it was. The only way H Property Management could really know this was if they had carried out an inspection. I was not satisfied that an inspection had been carried out properly or if an inspection was carried out to ascertain whether any signs of a breach of household occupancy, such as the number of beds, had been followed up. Mr K confirmed the shed was built eight or nine years ago and not intended for living accommodation.

A 14-day consultation letter was sent to the landlord and agent advising them of the hazards and most likely course of action. During this period, Maya and her family were rehoused by social services, after I provided them with my risk assessment. Yasmin was on housing benefit and a low income, claiming single-person discount for council tax. But it was sublet as an HMO and not in single occupancy. Yasmin's benefit was suspended pending investigation and eventually reinstated minus an amount owed for the period the property was operating as an unlicensed HMO.

I returned to the property in August and met Mr H the Director of H Property Management. I was able to verify that the shed was now empty of furnishings. The lock had now been changed, preventing access. I now felt it possible to serve a prohibition order as opposed to a suspended prohibition order and the outhouse would have an alternative use and not be reused as living accommodation.

A schedule of works in the order explored possible actions to reduce or mitigate the hazards identified, but irrespective of any improvements to the structure and upgrades of heating, insulation and provision of bathing and washing facilities. The cost of carrying out works to address excess cold, damp and mould, and inadequate sanitation and provision of water is prohibitive, but crucially it is conditional upon obtaining planning permission and building control approval, which was unlikely to be granted.

Aside from this, there is no possible way to ensure an appropriate means of escape to a place of ultimate safety in the event of a fire. The property is mid-terrace, the garden has no side alley and the only way out to the street is through the front door of the property, involving negotiating a route through the house and passing through the kitchen, which is a risk room for fire before accessing the hallway and front door. There is no practical way to overcome this.

Prohibiting this outbuilding for use as sleeping accommodation is also appropriate because there was a high likelihood that the outhouse could become reoccupied given the demand for housing in the borough.[1]

In making the decision to prohibit, I also considered sustainable development, and the definition set out in the Brundtland report, *'development that meets the needs of the present without compromising the ability of future generations to meet their own needs.'*[9] There is no way that a bed in a shed can possibly be seen to adequately meet the needs of the present or future if we were to entertain the possibility of remedying the hazards. If the main house was to be occupied by a single family, the only access to the shed is still through the hall and kitchen of the main terraced house and would be unacceptably intrusive to residents in the main house.

If the shed were to be subdivided to make a bathroom the available floor space would make for cramped accommodation with exceptionally poor conditions for any occupier. The garden is also small for the usage of two families in this way, and as a general rule, turning sheds into living space makes no provision for the parking of cars, cycles or for the storage of refuse. To apply such a policy to beds in shed across the borough would be unsustainable, and the long-term stability of the structure would be questionable given that it was not constructed with the benefit of any building control consent or supervision.

I also noted the backdoor kitchen glazing was still broken, and again raised the issue.

Mr H stated that he would not repair the glazing unless the tenants paid the cost and intended to evict Yasmin because he was no longer receiving rent. I was astonished by the lack of understanding; the occupiers' benefit had been suspended pending an investigation over the multiple occupancy, which had happened under the watch of Mr H who was supposedly managing the property.[1]

Civil financial penalties

The introduction of civil financial penalties (FPNs) under the Housing and Planning Act 2016 offered a potential change in our working practices against criminal landlords. This is an alternative to prosecution, allowing the local authority to impose fines of up to £30,000 for certain housing-related offences, including management and licensing offences,[10] higher than what had typically been handed down through the courts.[11]

Up until the beginning of 2015, prosecution fines were capped at £5,000 per offence, but a change in the law meant these fines were potentially unlimited.[12]

In my own cases against criminal landlords, the change did not always translate into the courts handing out a fine reflective of the severity of the offence. Accompanied by a journalist, my revisit to a property in Barking Road following a successful prosecution showed that even after being fined there is still no guarantee that the punishment will lead to the rectification of hazardous conditions.[13]

Through regulation 4 of The Rent Repayment Orders and Financial Penalties (Amounts Recovered) (England) Regulations 2017, the local housing authority can retain income received from civil financial penalties to meet costs incurred in carrying out enforcement functions.[14] An FPN can be issued in alternative to prosecution for certain offences under the act, including offences in relation to licensing of Houses in Multiple Occupation (section 72) and failure to comply with management regulations in respect of Houses in Multiple Occupation (section 234[4]).[3]

The existing selective licence was in breach of having more than the permitted number of households (one single family) as per the licence conditions (p. 31),[15] making it an offence under section 95(2) of the act.[3] The broken glazing to the backdoor of the property is an offence under the management regulations of HMO's section 234(3)[3] and regulation 7(2)(d) of the Management of Houses in Multiple Occupation (England) regulations 2006.[6]

It was decided that the application of FPNs was the most appropriate course of action, upon the licence holder and the managing agent, to address these issues. Although Yasmin and her husband had sublet the property, I did not feel it to be in the public interest to impose a financial penalty upon them, given that they had already had their benefit suspended.

In accord with the Department for Communities and Local Government's (DCLG) guidance, the local authority has to consider the facts of each case, weighing them against the following four dimensions or factors, before they can arrive at an appropriate amount.[10]

- Deterrence and prevention
- Removal of financial incentive
- Offence and History
- Harm to tenants

The DCLG guidance does not specify how to apply a monetary value to these factors, and local authorities are having to make their own decisions in this respect. Some local authorities like Newham use a scoring matrix created in an excel spreadsheet where the officer can place a score across each of the dimensions – lower scores for less serious aspects of the offence, and higher ones to reflect the severity.

The process is repeated across the four factors, and the total score is added at the end. The scoring is weighted to ensure it produces a total score capped at £30,000. The DCLG guidance considers the factor of harm to tenants to be the most important. The greater harm or the potential for harm, then the higher the amount of the fine should be.[10]

Deterrence and prevention

The guidance emphasises that the imposition of an FPN should be high enough to deter repeat offending and ensure the recipient does not reoffend. It should also be high enough to set an example and 'send a message.'[10]

I had to decide based on the facts how confident the local authority could be that a fine would act as a deterrent. In doing so, I considered the nature of the complaint which prompted the inspection as being a strong indicator of poor management, along with the lack of response to the original condition audit. The local authority could not be confident that the licence holder and his appointed agent were exercising proper management of the property.

If there is high confidence in this factor it leads to a low score. If there is low confidence, then it will lead to a higher score. In the case of the licence holder who has an agreement with H Property Management for them to manage the property, it was felt that the financial penalty would serve as a deterrent for repeat offending.

In the case of H Property Management, the appointed agent, the expectation is that an agent who manages properties would address issues of subletting and over-occupancy as a matter of course. The broken window, a more immediate and obvious defect, was ignored. There was low confidence that a penalty would deter repeat offending.

Removal of financial incentive

The guidance states the offender should not benefit as a result of committing an offence. It should not be cheaper to commit an offence than to ensure a property is well maintained and managed. The penalty has to be proportionate, set at a level that ensures there is a real economic impact on the offender demonstrating the consequences of failing to comply with the law and their responsibilities.[10]

To assess this aspect, the local authority should investigate the offender's assets, to determine if they have a large or small portfolio of rented properties and to gauge that asset value. The larger the portfolio, the higher the score. In the case of Mr K, licensing records showed that he is registered as the selective licence holder with other properties in the borough. I correlated the addresses with land registry searches, finding them all to be in Mr K's ownership. It was determined that Mr K has a large asset value, from which a large profit is yielded.

Bizarrely, the same could not be said for Mr H and H Property Management, who were a newly incorporated limited company that was yet to submit financial returns to companies' house. There were no other records of their appointment as managing agents for other licence holders or in their own right. The company's web presence for renting or selling property was minimal at the time the calculations were made, so they were deemed to have little asset value. As it would later transpire, this aspect was underscored, but you can only go with the information you have available to you.[1]

Offence and history

Here an assessment is made of the culpability of the offender, and it is subject to the evidence acquired by the local authority in the course of their investigations. Account is taken of previous enforcement action taken against the offender by the local authority, that is, previous prosecutions; FPNs or statutory notices served; the outcomes, compliance or failure to comply with notices; and history of notices being served for similar offences, all help to determine the extent of any history of poor management.

A higher score is applied for multiple offenders with long histories of enforcement. Where there is no previous history of enforcement then the lowest score is applied. However, between these factors depending on the individual case, if there is evidence that the offender committed the offence deliberately and knew or ought to have known they were in breach of their responsibilities (this could be because they the local authority have acted against them before for the same or for similar offences) then this should also be a factor in how low or high a score is applied.

I took account of the prohibition order I had served owing to the breach of permitted households resulting in Maya and family being exposed to category 1 and 2 hazards.[1]

Harm to tenants

The DCLG regard this as a very important factor. The greater the harm or potential for harm (this may be as perceived by tenants) the higher the amount should be when imposing an FPN. This could be where a defect presents a hazard to the health of the occupier if not remedied; consideration is also given to the tenant's stress or fear of possible injury arising from the hazard.[10]

The greater the risk or potential for harm then the higher the score. Maya and her husband provided information on the impact of living in the shed, where it was cold and exposed to penetrating damp. The occupiers were worried for the health of their baby, and the stress was exacerbated by the experience of being trapped in the garden.[1]

Notice of intention to issue a financial penalty

The licence condition breach of exceeding the permitted number of households is what led to the property becoming an HMO and the offence of failing to licence an HMO. In a way this offence is the same, and just as you couldn't be charged in court twice for the same offence you cannot fine the licence holder twice for the same offence.

H Property Management are not the licence holder so cannot be fined for the licence condition breach under section 95(2) of the act.[3] But both Mr K and H Property Management are persons managing and in control of the property as per the definition under section 263,[1] because they both receive the rent. For this reason, both parties were considered liable for failing to licence the HMO and for the management regulation breach in relation to the kitchen window. It made sense to fine them both for these offences.

If there are multiple offenders involved in a case where there is both a landlord and managing agent who are both culpable of the offence, then fines will be calculated per offender using the financial penalty scoring criteria. It may be that this results in differing levels of fines being applied to the offenders, and this will have been arrived at by weighing the circumstances of the individual offender against the four factors. For example, if one offender has a previous history of enforcement against them, and the other does not, then the offender with the previous history will have a higher score than that of the other.

The same would also be true if one offender is shown to have a larger portfolio than the other, as was the case with Mr K and H Property Management. Mr K's fine amounted to £10,000 for both offences and H Property Management were fined a total of £5,000 for both the offences.[1]

The right to make written representations

A recipient of the notice of intention has the right to make written representations to the local authority over a 28-day period.[10] I call it trial by post.

A key point during representations was H Property Managements' attempt to distance themselves by claiming the firm don't manage the property but provide 'rent collection service only,' despite the fact that Mr H had agreed to his firm being appointed as managing agents on the property licence.[1]

I asked to see the contract between both parties. The terms of business confirmed rent collection was by standing order. A section in the contract headed 'Full Management,' outlined additional services provided, including the dealing of 'day-to-day management' as well as minor repairs for a fee agreed by the landlord, except in the case of emergency or to enable the landlord to comply with a statute.

There were further clauses to the effect that the agency can instruct contractors and tradesmen on the landlord's behalf and deduct the cost of repair and maintenance from rent. With regard to property inspection, the agency agreed to undertake these three times per year. The documents were signed by both parties, and I was satisfied that the terms consisted of an agreement for Mr H and his firm to manage the property.[1]

The representations were not accepted, because rent collection would still make the agency persons managing or in control of the HMO in accord with section 263 of the act.[3] After 28 days, I wrote to them both with my decision to issue final FPNs. Shortly after that, I received notification from the First Tier Tribunal Residential Property Chamber (FTT) that the parties were appealing the fines, and a hearing date was set.[1]

The tribunal

At the FTT, the tribunal dismissed appeals against both Mr K and H Property Management Ltd, for the management regulation breach and confirmed the financial penalties. They dismissed the appeal by H Property Management Ltd for failure to licence the HMO and confirmed the penalty, but in the case of the licence holder the appeal was allowed, as his reasonable excuse that the agent was at fault was accepted.[16]

Overall, I was satisfied with the outcome, as this was one of the first of our FPN cases to go before the tribunal. We were keen to see how the tribunal would view the matrix and the methodology behind it because many cases involve both licence holders and agents. We wanted to see how the tribunal would judge the division of responsibility between them.

The facts in my statement were largely uncontested, and the tribunal found that the criterion used to determine the level of fines was 'properly based on DCLG guidance' and were satisfied that the penalties were appropriate.[16]

The defence tried to argue that ASB in the property was the fault of the 'shed occupiers,' but I argued that allowing over-occupancy was antisocial. The breaking of the window was used by the defence to illustrate their point; however, I disagreed very strongly. Maya and her family had been placed in unacceptable living conditions with no access to bathroom facilities. With proper management in place this would have been prevented, as would the danger of living with a damaged backdoor, an invitation for entry by intruders providing a risk to the family in the house.[16]

Our case centred on the fact that either the property was not properly inspected or that the appellants knew of the subletting and chose to ignore it. We accepted that the tenants had sublet the property without authorisation but the persons managing and in control had allowed it to continue. Subletting by tenants is not an acceptable defence for persons managing or subject to licence conditions. I was pleased that the tribunal largely agreed with me, and placed weight on the harm to tenants, as this often gets overlooked in court cases.[16]

The notification to impose a financial penalty had the effect of making the defendants address a management regulation issue they were otherwise ignoring. They eventually paid to fit a new glass panel in the door. It is my belief based on experience, that even threat of prosecution would not have instilled a change in this attitude. Together with the prohibition of the outhouse, the intervention had worked to remove an uninhabitable and hazardous structure from exposing vulnerable occupiers to risk.

The ruling has meant that Mr K, Mr H and H Property Management Ltd are persons of concern and all associated licences of theirs are issued for only one year, and they are now subject to heavier regulation. Whilst Newham still carry out prosecutions, this case helped to bring a change in working policy to adopt civil FPNs as the primary intervention for addressing housing offences caused by criminal landlords.[16]

References

1. Oatt P, An examination of effective enforcement methods within licensing of private rented housing in the London Borough of Newham, for tackling the worst effects of poor housing 2018 Case Study submitted to CIEH. August 2018.

2. London Borough of Newham. Private rented property licensing guide for landlords and managing agents [online]. (June 2012). Available from: www.Newham.gov.uk/Documents/Housing/PropertyLicensing-GuideLandlordsAndManagingAgents.pdf [Accessed 1st August 2018.]

3. legislation.gov.uk. *Housing Act 2004* [online]. Available from: www.legislation.gov.uk/ukpga/2004/34/contents [Accessed 27th February 2016.]

4. Mishkin P, Moffatt R. A review of multi-agency enforcement and discretionary property licensing to tackle Newham's private rented sector. In Stewart J (ed.) *Effective Strategies and Interventions: Environmental Health and the Private Housing Sector.* London, England: Chartered Institute of Environmental Health, 2013, pp. 12–14.

5. Oatt P.J. Exploded Isometric property floorplan. (2019) Phoenix Oatt. Graphic illustration.

6. legislation.gov.uk. *The management of houses in multiple occupation (England) regulations 2006* [online]. Available from: www.legislation.gov.uk/uksi/2006/372/pdfs/uksi_20060372_en.pdf [Accessed 31st August 2019.]

7. Office of the Deputy Prime Minister. Housing health and safety rating system guidance (Version 2). ODPM Publications, West Yorkshire, 2004.

8. Gov.UK. Police and Criminal Evidence Act 1984 (PACE) codes of practice [online]. Available from: www.gov.uk/guidance/police-and-criminal-evidence-act-1984-pace-codes-of-practice [Accessed 1st September 2019.]

9. Keeble B.R. The Brundtland report: 'Our common future'. Medicine and War, 1988; 4(1): 17–25. doi:10.1080/07488008808408783

10. Ministry of Housing Communities and Local Government. Civil penalties under the Housing and Planning Act 2016. Crown Copyright [online]. (2018). Available from: https://assets.publishing.service.gov.uk/government/uploads/system/uploads/attachment_data/file/697644/Civil_penalty_guidance.pdf [Accessed 1st August 2018.]

11. Brown J. *Rogue landlords 'should face tougher penalties' for squalid conditions* Independent [online]. (14th June 2014). Available from: www.independent.co.uk/news/uk/politics/rogue-landlords-should-face-tougher-penalties-for-squalid-conditions-9536243.html [Accessed 22nd August 2019.]

12. Local Government Association. *Prosecuting landlords for poor property conditions.* Private sector housing research, June 2014.

13. Booth R. *'I clean for Chelsea FC and live in squalor': inside illegal housing.* The Guardian [online]. (29th January 2018). Available from: www.theguardian.com/society/2018/jan/29/chelsea-fc-squalor-uk-illegal-housing-london [Accessed 22nd August 2019.]

14. legislation.gov.uk. *The rent repayment orders and financial penalties (amounts recovered) (England) regulations 2017* [online]. Available from: www.legislation.gov.uk/uksi/2017/367/regulation/4/made [Accessed 31st August 2019.]

15. London Borough of Newham. Private rented property licensing guide for landlords and managing agents [online]. (June 2012). Available from: www.Newham.gov.uk/Documents/Housing/PropertyLicensing-GuideLandlordsAndManagingAgents.pdf [Accessed 21st December 2016.]

16. LON/00BB/HNA/2017/0018 & 19. Residential property tribunal decisions. First-Tier tribunal property chamber (residential property) [online]. (1st March 2019). Available from: https://assets.publishing.service.gov.uk/media/5c7916d2ed915d29eb6a0056/LON00BBHNA20170018___19_-_Bristol_Road.pdf?_ga=2.251520774.1464556537.1566800783-304649890.1566047204 [Accessed 23rd August 2019.]

4   HHSRS or licensing?

What are the different aims and objectives of the regulations set out in parts 1, 2 and 3 of the act? Why do we need licensing regulation as well as notices to improve or prohibit? And if we need to license properties, why have they been separated out into licensing of mandatory and additional House in Multiple Occupation (HMOs) and selective licensing?

Under part 1, the local authority has a duty to act when category 1 hazards are identified, and may act where there are category 2 hazards.[1] There is a period of negotiation often met with refusal, sometimes a bit of foot-stomping and crocodile tears, or threats to have you sacked. Once the consultation is over, and you have been assured of your ongoing employment, the authority pursues their course of action. Section 5(3–4) is clear that if only one course of action is available to the authority in relation to the hazard they must take that course of action. If there is more than one course of action, then the local authority opts for the one considered to be the most appropriate.[1] The case study covers how this was applied to the decision regarding the shed.

If the chosen course of action proves unsatisfactory, the local authority still has a duty to address the hazard. The choice here is to either repeat the same course of action or take another action from the available options, sections 5(5) and 7(3).[1] In the case of the shed, the prohibition order was the best course of action; if anyone were to occupy it again Mr K and his agent would now be looking at a prosecution for breach of the order.

An improvement notice is served for category 1 or 2 hazards or a combination of both, section 12(5).[1] The recipient of the notice is required to take remedial action, and the notice will outline the hazards and the works required. The act states under section 11(5) that at minimum, remedial action must ensure the hazard ceases to be a category 1 hazard, so in the best case it is eliminated altogether or it is reduced.[1]

This is where you can get into problems; damp proof works are a classic. You can be very prescriptive about the type of work required to deal with a serious case of rising damp in a pre-1920s property, and a reluctant landlord will do some basic cosmetic work to clean off mould, re-plaster the interior affected areas but fail to address the root cause, or partially address it through chemical injection of the damp-proof course, totally ignoring the improvement notice requirements to cut a trench and reduce ground level to allow provision of air grids and sub-ventilation, repair brickwork or render the exterior, and rearrange the rainwater pipes so they discharge below the gulley and not all over the brickwork. Basic steps that will ensure the problem does not recur.

In this scenario, the landlord has not eliminated the hazard altogether; the local authority then has to determine if work is sufficient to reduce the category 1 hazard. Then the question is whether or not to pursue the landlord through the courts for failing to comply with the outstanding requirements. For any offence under the act, the defendant has available to them a reasonable excuse defence for not complying with the regulation.[1] Or serve yet another notice.

There is very little set down in law defining the reasonable excuse defence. Section 101 of the Magistrates Courts Act states that the burden of proof for a defendant's reliance on an excuse rests with the defendant.[2] The local authority must be satisfied of guilt beyond a reasonable doubt. If a landlord does nothing, then proving guilt is easier, but the cases become murky where a half-hearted attempt has been made to comply.

The length of time for cases to get to court is also a problem, and, in that time, the landlord may go ahead and do more works and try and mitigate in the hope of getting let off or handed a lower fine.[3]

Another source of delay with notices is an appeal to the First Tier Tribunal Residential Property Chamber (FTT). The recipient has a right to appeal before the operative date, sometimes on frivolous grounds but enough to get the appeal allowed and the notice suspended, prolonging the occupier's exposure to hazardous health conditions.

When Newham made the case for designating the whole borough for licensing, an equality impact assessment (EQIA)[4] was made as part of a 2012 cabinet report, and amongst the reasons for the designation was that

> the standard enforcement regime can be complicated, time-consuming and expensive which makes it difficult for local authorities to act quickly against poorly managed and maintained private rented properties.

The cabinet report stated that the difficulties with regulating the private rented sector were increasing due to the illegal use of out-buildings as accommodation making Newham a 'hotspot' for such activities, where evasive landlords did not comply with legislation. According to the report, a landlord survey carried out in 2009 showed that over 50% of Newham landlords were failing to carry out credit checks on tenants, 41.3% did not provide tenants with information packs and 22.9% did not provide tenants with emergency contact details. As much as 21.9% did not use inventory reports or schedules, and 71.2% had no trade waste agreement.[5]

Section 56 of the act makes clear that justification of an HMO licensing designation must be in consideration that a 'significant proportion of the [areas] HMOs .... are being managed sufficiently ineffectively [giving rise to] particular problems,' for occupiers and members of the public. Section 57 makes clear that the local authority should adopt a coordinated approach to combine licensing with other available courses of action.[1]

Schedule 4 of the act specifies mandatory conditions for HMO and selective licenses which must include requirements for smoke alarms, gas and electrical safety, and production of associated documentation on demand from the local authority.[1]

The mandatory and discretionary licence conditions available to local authorities would not be sufficient for addressing hazards arising from disrepair. Let us say, for example, there were loose or missing tiles in a kitchen and damage to work surfaces, in a selective licensed property. If a local authority issued a licence with conditions to address these defects, and a routine follow-up inspection found disrepair in the kitchen, could it be said with any conviction that it is the same set of tiles or the same worktop? Depending on the passage of time, a landlord could argue that the original matters were addressed and these matters are a new occurrence.

The Court of Appeal recently had an opportunity to examine the different approaches in *Brown v. Hyndburn Borough Council* where Brown appealed two discretionary selective licence conditions imposed by the council: first, requiring carbon monoxide (CO) detectors to be fitted and maintained in properties with gas, and second, to ensure that over the licence period the property has an Electrical Installation Condition Report. If that report showed any unsatisfactory items, the condition required they be remedied within 38 days. Brown argued against local authorities using licence conditions to require improvements or upgrades or provide new equipment.[6]

The court found that both conditions could not be imposed by Hyndburn because they are contrary to the discretionary conditions under section 90(1) of the act,[1] which regulate the management use or occupation of the house. Both conditions relate to improvement and upgrading of electrical installation and fitting of new facilities, that is, CO detectors.[6]

The court compared the discretionary powers for selective licences to those set out for HMOs. Sections 64, 65 and 67 of the act make clear that an HMO licence may include appropriate conditions to address antisocial behaviour (ASB), overcrowding and regulate the standards and quality of bathrooms, toilets, washing, bathing, laundry facilities and kitchen areas for cooking, storage and food preparation.[1] So, this covers regulation of management, use and occupation of the HMO, and its conditions and contents.

Section 90 makes clear that selective licensing may include appropriate conditions to restrict occupation and address ASB. It then goes on to state that it may include conditions to keep facilities in repair and proper working order, but states under section 90(5) that it must use the act's part 1 powers to identify and remove or reduce category 1 and 2 hazards.[1]

The court also pointed out that this does not '*connote that a power is conferred*' to regulate the conditions and contents of a property.[6]

For HMOs, section 55(5)(c) of the act makes clear that the local authority must make sure there are no issues that need to be addressed under part 1 in HMOs.[1] Considering the separate functions under part 1 of the act to deal with hazards, the court found that the distinction between the discretionary conditions under parts 2 and 3 must have been intentional.[6]

HMO landlords are under a duty under management regulations stipulated via section 6(3)(a) to ensure that every fixed electrical installation is inspected and tested at intervals not exceeding five years by a person qualified to undertake such inspection and testing.[7] It is not unreasonable to request a document that they are already required to have in law. For other types of private rented properties, the government announced in January that they intend to introduce compulsory five-yearly electrical installation checks as soon as parliamentary time allows.[8] A licensing condition to this effect would therefore be appropriate once this legislation is enacted.

For the most part, inspections under part 1 of the act are largely carried out in response to a tenant's complaint. The 2015 Battersby report showed a decrease in rate between complaints and local authority inspections from 2011 onwards. The report suggests it is the lack of resources that led to leaving many complaints unaddressed through council staff either gatekeeping or signposting complainants to other services. Additionally, pressure to find temporary accommodation for increasing rates of homelessness displacements is considered along with conflicting objectives between departments overseeing temporary allocations and housing teams responsible for enforcement, thus identifying a weak link between agencies that should be working in partnership.[9]

One of the objectives of Newham's licensing scheme is that

> All landlords and managing agents operate at the same minimum level of professional standards[10]

To achieve the aims, the licensing designations under parts 2 and 3 of the act are regulated through multi-agency collaboration and a proactive schedule of unannounced visits across the borough.

The functions under part 1 are traditionally administered as a reactive model in response to complaints made by occupiers, through prearranged inspections. Before licensing, Health and Housing Safety Rating System (HHSRS) was the main tool used by the local authority to regulate the private rented sector. But the tool was ineffective because it deals with houses on a case-by-case basis, as and when problems are bought to the local authority's attention. It does not help to drive up professional standards across the borough. The tool is used to encourage a landlord to act, and often this results in achieving only the minimum standard, reducing a category 1 hazard to category 2 and not eliminating it altogether.

For a portfolio landlord with no interest in operating to professional standards, it makes perfect sense to sit and wait until they are inspected and keep standards at their lowest point, through a perverse form of cost–benefit analysis. Following a part 1 inspection, the landlord effectively gets a free survey, complete with a schedule of works, an idiot's guide if you will, to carrying out remedial actions that they should be doing anyway under general maintenance. Those that take the cost–benefit analysis further and weigh up the amount of rental income they received against the scale 5 fine they risked incurring for noncompliance with the notice, which used to be up to £5,000, would consider the risk beneficial because it outweighed the cost.[11]

Whilst section 49 of the act allows for local authorities to charge for the service of notices under part 1 for assessing whether or not to take the action and the actual serving of the notice, the average charge according to the LGA for service of an improvement notice is £360 (p. 19),[12] very small in proportion to the amount of officer's time taken to make all the background checks, serve section 239 notifications, inspect, risk assess and go through a consultation period before drafting and serving the notice.

The prosecution fine is now unlimited, and whilst fines have increased above £5,000 they are not always punitive or reflective of the risk of harm. Low fines were recorded in two recent prosecutions of landlords: one by Thurrock, where the property had exposed live wiring, dampness and several other category 1 hazards, and another in Harlow where the landlord had ignored notices over a dangerous structure, which resulted in fines amounting to £7,000.[13,14] An HMO landlord in Oldham earlier this year charged with serious fire and electrical safety breaches was fined less than £6,000.[15] If part 1 is the only tool used in a local authority with a large private rented sector and an increasing number of HMOs, then it is a race to the bottom, the application of a sticking plaster to a cancerous disease.

The HHSRS tool provides a risk assessment that is evidence based, but the evidence base is now quite old. Each time a local authority takes an informal action over a category 1 hazard instead of an evidence risk-based approach, it diminishes that data even further. It is a bit of a circular argument to say that the data on which the HHSRS is based are out of date, but they are out of date because local authorities have failed to contribute sufficiently to update the data with practical examples. The recent government review of HHSRS has borne this out, finding that in order to improve and modernise HHSRS assessment the enforcement guidance needs to be reviewed and updated and a comprehensive set of worked examples be developed incorporating new standards into the HHSRS process.[16]

Licensing conditions obligate landlords to manage their own properties or face consequences. Licensing provides local authorities with a means to quantify the private rented sector that they regulate, and determine how many rented properties are in the borough and of what type, and also to distinguish between compliant and noncompliant landlords, because criminal landlords are conspicuous by their absence, for example, their failure to submit licensing applications.

The licensing scheme provides a regulatory framework to identify and target criminal landlords at scale in a way that part 1 regulation cannot, enabling local authorities to deploy their resources to inspect, identify noncompliance and enforce on matters that would otherwise go unnoticed. Because the licensing parts of the act (sections 2 and 3) say that local authorities must use part 1 to address category 1 hazards, then this in turn should be the driving force for local authorities to work out how to use these separate arms in combination.

It was a tenant complaint that brought the shed in Forest Gate to my attention. Rather than immediately respond by inspecting under part 1, which would involve giving 24 hours' notice to the landlord under section 239(5) of the act,[1] I was able to enter unannounced under section 239(7) instead, identify the HMO-related issues and gather sufficient evidence to return under section 239(5) and take part 1 action in combination with licensing enforcement.

References

1. legislation.gov.uk. *Housing Act 2004* [online]. Available from: www.legislation.gov.uk/ukpga/2004/34/contents [Accessed 27th February 2016.]
2. Legislation.gov.uk. *Magistrates' Courts Act 1980* [online]. Available from: www.legislation.gov.uk/ukpga/1980/43/section/101 [Accessed 4th September 2019.]
3. Brown, J *Rogue landlords 'should face tougher penalties' for squalid conditions* Independent [online]. (14th June 2014). Available from: www.independent.co.uk/news/uk/politics/rogue-landlords-should-face-tougher-penalties-for-squalid-conditions-9536243.html [Accessed 22nd August 2019.]
4. London Borough of Newham. Borough wide landlord licensing (BWLL) EQIA Template. (1st June 2012). [online]. Available from: www.Newham.gov.uk/Documents/Council%20and%20Democracy/BoroughWide LandlordLicensingEqIA.doc [Accessed 4th September 2019.]
5. Dick I. London Borough of Newham Cabinet Report, Private rented property licensing. (June 2012). [online]. www.landlords.org.uk/sites/default/files/$cabinetreportfinal.doca_.ps_.pdf [Accessed 4th September 2019.]
6. British and Irish Legal Information Institute. Brown v Hyndburn Borough Council (2018) [2018] WLR 4518, [2018] LLR 239, [2018] HLR 19, [2018] EWCA Civ 242, [2018] 1 WLR 4518, [2018] WLR(D) 115 [online]. Available from: www.bailii.org/ew/cases/EWCA/Civ/2018/242.html [Accessed 4th September 2019.]
7. legislation.gov.uk. *The management of houses in multiple occupation (England) regulations 2006* [online]. Available from: www.legislation.gov.uk/uksi/2006/372/pdfs/uksi_20060372_en.pdf [Accessed 31st August 2019.]
8. Ministry of Housing Communities and Local Government. *Housing Minister tightens up rules on electrical safety to better protect renters* [online]. (29th January 2019). Available from: www.gov.uk/government/news/housing-minister-tightens-up-rules-on-electrical-safety-to-better-protect-renters [Accessed 21st September 2019.]
9. Battersby S. *The challenge of tackling unsafe and unhealthy housing* [online]. (December 2015). Available from: http://sabattersby.co.uk/documents/KBReport2.pdf [Accessed 27th February 2016.]
10. London Borough of Newham. Private rented property licensing guide for landlords and managing agents [online]. (June 2012). Available from: www.Newham.gov.uk/Documents/Housing/PropertyLicensingGuide-LandlordsAndManagingAgents.pdf [Accessed 21st December 2016.]

11. Brown J. *Rogue landlords 'should face tougher penalties' for squalid conditions* Independent [online]. (14th June 2014). Available from: www.independent. co.uk/news/uk/politics/rogue-landlords-should-face-tougher-penalties-for-squalid-conditions-9536243.html [Accessed 22nd August 2019.]

12. Local Government Association. *Prosecuting landlords for poor property conditions.* Private sector housing research, June 2014.

13. Thurrock Gazette. *Landlord fined £7,000 after failing to provide adequate living conditions* [online]. (18th August 2019). Available from: www. thurrockgazette.co.uk/news/17834684.landlord-fined-7-000-failing-provide-adequate-living-conditions/ [Accessed 22nd August 2019.]

14. yourharlow.com. [online]. (17th August 2019). Available from: www. yourharlow.com/2019/08/17/pytt-field-landlord-fined-6000-due-to-poor-condition-of-house/ [Accessed 22nd August 2019.]

15. Jackson N. *Oldham landlord fined nearly £6,000 for safety breaches.* Oldham Times [online]. (19th March 2019). Available from: www.theoldhamtimes. co.uk/news/17511098.oldham-landlord-fined-nearly-6000-for-safety-breaches/ [Accessed 22nd August 2019.]

16. Ministry of Housing Communities and Local Government. *Outcomes of report on Housing Health and Safety Rating System (HHSRS) scoping review* [online]. (11th July 2019). Available from: www.gov.uk/government/publications/housing-health-and-safety-rating-system-outcomes-of-the-scoping-review/outcomes-of-report-on-housing-health-and-safety-rating-system-hhsrs-scoping-review [Accessed 22nd August 2019.]

A shed (interior) being used as a bedroom.

Overcrowded bedroom in an HMO.

Unsafe electrical wiring seen in a kitchen.

Exposed void in a kitchen ceiling, unable to contain or limit the spread of fire.

A subdivided room in an HMO with no window.

Damage to ceiling from a roof leak, and carboard being used to block broken or missing panes of glass in windows.

Staircase with missing spindles.

Garden at rear of HMO.

5 The policy analysis

A policy analysis was carried out to evaluate Newham's first five-year licensing designation at the midpoint of its scheme, using Morestin's 'Framework for Analyzing Public Policies.'[1,2] The framework is ideal for policy analysis when there is no sole encompassing document and can be applied against different evidential sources accounting for policy context, justifications and supporting data.

Morestin's framework uses a structured analysis approach developed from a public health perspective under the axis of effects and implementation across the six dimensions of effectiveness, unintended effects, equity, cost, feasibility and acceptability, and enables a summary analysis of the advantages and limitations of the policy under scrutiny.[1,2]

Framework questions are indicative, and it is not always possible to encapsulate all elements associated with each dimension; consequently, there may be gaps in the end product. Morestin cautions on making causal relationships in public health policy where policy is one of many factors, but stresses the value of looking at intermediate effects.[1,2]

Effectiveness

The policy under analysis is not contained within a single document. A cabinet report dated June 2012[3] outlines the mapping of service actions and priorities around applicant support, license applications, inspections and enforcement, investment and training of extra staff, development of enforcement policies and agency intelligence sharing (inputs).

The effectiveness of the project can be seen through the outputs at the midpoint of the scheme's implementation in Newham's proposed renewal for 2018. Licensing allowed Newham to learn more about its private rented sector stock, landlords' ownership patterns and portfolio sizes than previously perceived. Quantitative data on licensing show in 2016 there were over 25,000 landlords and 38,000 licensed properties.[4]

Inputs

Section 237 of the act[5] allows sharing of council tax and housing benefit data. Overcrowding assessments are made under the crowding and space standards set out in part 10 of the largely repealed Housing Act of 1985,[6] enabled within section 216 of the current act.[5] This measurement criterion determines permitted occupancy levels by room sizes, available bathrooms and kitchens and is used to create the licence conditions for that particular property.[7]

Outputs

Newham working in strong partnership and carrying out joint operations with other agencies such as the police, fire brigade and HM Revenue and Customs (HMRC) has the shared objective of investigating housing-related crimes. At the end of 2016, this was shown to have resulted in 368 multi-agency operations leading to 3,555 unannounced visits, 908 landlord prosecutions, 2,170 notices served to improve housing conditions and 652 arrests made by police and immigration.[4]

Outcomes/impacts

Independent research carried between 2011 and 2014 shows Newham's prosecution rate exceeded that of all other London boroughs combined, demonstrating the effectiveness of its collaborative enforcement policy, amidst increasing annual prosecution rates, indicating that the culture of noncompliance remains endemic.[8]

The number of prosecutions when balanced against the number of licenses issued indicates there is broad compliance amongst Newham's landlords, but that this culture of noncompliance rests only within a proportion of landlords who operate outside of the law.

Unintended effects

Newham anticipated the possibility that criminal landlords would retaliate by illegally evicting tenants and that this activity would increase in response to licensing. Collaboration with the police, Shelter and other third-party organisations disseminates awareness amongst partner agencies of the tenants' rights to protection from harassment and illegal eviction.[9]

Equity

The groups most affected through borough-wide licensing regulation are landlords and their tenants. According to 2011 census data, Newham's average household size was 3.3 persons per household,[10] and the number of arrests (652) shown in the outputs arising from a potential 11,731 households equates to 5.56% arrests amongst the total number of potential households:

$$652 / (3,555 \times 3.3) \times 100 = 5.56\%.$$

Whilst it is a large amount of arrests, relative to the population of the private rented sector it shows the majority of tenants were found to be law-abiding potential witnesses to housing crimes.

For compliant landlords, there is a 'light touch' approach, whereas noncompliant landlords face enforcement.[3,7] Prosecution is considered a first course of action for significant management failures; prolific offenders risk losing their licenses, but guidance is provided to facilitate landlords in meeting regulatory obligations.[7]

Adverse short-term consequences to landlords adjusting to new regulatory burdens and fees were considered against long-term impacts upon market competition becoming more equitable as standards are driven up because noncompliant landlords circumventing rules and undercutting through charging individuals lower than market rents (by overcrowding and charging room rates) will no longer have a commercial advantage.[8] By the end of 2017, 28 Newham landlords were banned from holding licenses.[11]

Data used in Newham's core plan draw on the Indices of Multiple Deprivation (IMD) for 2007 showing the borough has persistently high levels of unemployment and low skills. Household income is significantly lower than the London average; consequently, Newham has high levels of private rents supported by housing benefit payments.[12,13]

Borough-wide licensing cannot address employment and low incomes, but it enables early intervention through enforcement and education of private rented sector landlords renting to vulnerable tenants. Licence conditions direct landlords to prevent overcrowding by abiding with the prescribed occupancy levels of their licence, based on room sizes and amenities.[7] Abiding by the standard property licence conditions on tenancy management corrects the inequality between the landlord and tenant relationship whereby tenancy agreements, deposit protection, better maintenance, regular inspections and measures to deal with antisocial behaviour (ASB) are introduced into managerial practices across all licensed private rented sector properties.[7]

The policy also enables Newham to reinforce liability for unpaid council tax upon the landlords of Houses in Multiple Occupation (HMOs) instead of the transient occupiers.[14.] This is evidenced through council tax data gathered ahead of inspection by powers under section 237 of the act,[5] suggestive of multi-occupancy that is later confirmed through inspection.[15]

Table 5.1 Axes of Effects Analysed Using the Dimensions of Effectiveness, Unintended Effects and Equity

Axis and Dimensions		Inputs	Outputs	Outcomes
Effects	Effectiveness	Employment of extra admin and inspection staff for: Applicant support Inspection/ regulation Enforcement/legal Training to provide knowledge and skills Enforcement policy Agency referral – police, council tax, ASB teams Processing of license applications Online license application service integrated with council data systems	38,000 licenses issued (35,799 predicted licensable population based on 2011 census) 4,219 licensed HMOs 368 multi-agency enforcement operations 3,555 unannounced visits 2,414 HMOs and 6,000 other properties monitored for council tax arrears (year to date)	25,000 landlords as licence holders 908 prosecutions (cases sent to legal services by crime) 2,170 enforcement notices served 28 landlords banned from holding a licence 652 arrests during multi-agency enforcement operations Council tax recovery (see cost table 4)
	Unintended effects	Increased retaliatory and illegal evictions	Prosecution for illegal eviction/ harassment Multi-agency working to counter illegal eviction	Reduction of illegal evictions/tenancy harassment

…is and Dimensions	*Inputs*	*Outputs*	*Outcomes*
Equity	Persons most affected: Landlords Bad landlords undercut good landlords charging less rent per person to rent overcrowded HMOs Tenants exploited	Light-touch approach for compliant landlords Prosecution of noncompliant landlords	Short-term consequences of greater regulation on landlords Long-term behaviour change as standards are driven up Competition among landlords becomes more equitable altering market conditions so that they are no longer advantageous to criminal landlords

Implementation

Cost

The net cost of the scheme for 2012/2013 was £329,000, offset against the scheme costs for 2013 to 2014.[3] Prior to the beginning of the licensing designation, 26,000 applications were made taking advantage of an early discount fee of £150 per application, equating to an approximate revenue of £3.9 million.[16]

Staff costs are financially supported through the fee income of the first five years. Recovery of legal fee costs of enforcement is sought from the courts following successful prosecution. In comparison to the scheme, Newham has a large safety and enforcement division, spending £17.5 million annually addressing ASB and enviro-crime.[3]

Under council tax legislation HMO landlords are liable for tax and not the occupiers.[14]

In a recent report to Local Government Association (LGA) on Council Tax fraud in Newham, Quinn[17] reported that private rental arrears were highest in HMOs due to the transient nature of tenure and fraudulent behaviour of landlords. The '*known*' 4,219 licensed HMOs had a combined arrears amount of £930,373. In addition, Newham suspected there were up to 6,000 hidden HMOs incorrectly licensed as selective single-family dwellings with a combined arrears of £1,761,774.

Council tax officers collaborating with housing enforcement combined knowledge, gathered evidence and chased the debts, which resulted in a recovery of over £5 million in council tax arrears from HMO landlords,[18] thus addressing another inequality between compliant and noncompliant landlords making a profit through circumventing the law. If the scheme were not introduced, the tool of licensing conditions would not be available to apply across the borough.[3]

Feasibility

Given that the scheme continues, its feasibility is self-evident. But reasons why this is so are largely because Newham made good use of its existing infrastructure. A large team of Metropolitan Police Officers were already based within the council as part of an existing enforcement and safety team.[3]

This resource is used to collaborate with council officers on borough-wide licensing multi-agency inspections. The fact that officers already have established working patterns with other Newham departments is indicative that there is clarity over the roles in partnership.[3]

Acceptability

Newham policymakers consider licensing to be an instrument that helps reduce transience in the borough,[10] introducing stability to the sector, because the temporality of occupancy means that residents are less socially invested. The core plan argues that high levels of migration together with low rents means that many people pass through without '*setting down roots.*' The low rents are made possible through the number of flat or HMO conversions of family dwellings.[12] Studies into Newham's population growth show that it grew by 13.7% between 2007 and 2011, and whilst there are large levels of outward migration, Newham's population increased from 240,000 in 2012 to 318,000 in 2015.[19-21]

Stakeholders were concerned about the costs of dealing with fly tipping in the borough. Between 2014 and 2015, Newham spent £3.4 million cleaning up waste, more than any other London borough.[22] Transiency in the private rented sector is a large contributory factor;

licence conditions address transiency because landlords can be prosecuted if refuse is not stored appropriately and for not providing new occupiers with tenancy agreements.[7]

Licensing has a form of engagement built into its development process by law under sections 56 and 80 of the act,[5] obliging local authorities to undertake area-wide consultations and stakeholder engagement and consider representations without which government can refuse permission to designate licensing. The residential survey was a stratified sample of 415 people interviewed face to face by market research consultants Opinion Research Services (ORS),[10] finding that 80% of those surveyed were in support of the licence conditions. ORS oversaw the formal consultation between February and April 2012, additionally carrying out an online stakeholder questionnaire, two deliberative forums with landlords and agents as well as one with private rented sector tenants.[10]

The National Landlords Association (NLA) and the Residential Landlords Association (RLA) representing a large proportion of landlord members responded at length to the stakeholder questionnaire. The NLA response of May 2012 concedes that Newham has 'substandard and poorly managed properties', proposing an alliance between Newham and good landlords to create a programme of landlord accreditation, combined with financial incentives for landlords to join[23]; however, previous attempts to incentivise uptake of free membership to Newham's landlord accreditation scheme had met with little response.[3]

The RLA response criticised the licensing conditions for being excessive,[24] a view shared by 47% of private landlords.[10] The licence conditions are based upon criteria that are legally required to be incorporated into licensing schemes under the act, schedule 4, sections 67 and 90.[5]

The RLA also stated that Newham failed to consider taking a staged approach to deal with the worst areas first on a case-by-case basis[24]; however, the Equality Impact Assessment (EQIA) clarifies that problems in the private rented sector with ASB and poor management is distributed across all wards and not exclusive to particular property types, and for this reason the designations are borough-wide.[9]

Landlords concerned about costs and regulatory burdens requested that Newham's enforcement approach should be 'light touch' towards landlords with good business practises. In response to this, Newham provided an online license processing and payment option, with a simple fee structure and discount fee period.[3]

Durability

Prior to the roll out of licensing it was believed that Newham's private rented sector consisted of only 5,000 landlords.[4] But when Newham's 2011 census data for tenure and dwelling type[18] was analysed, it established that there were 33,118 properties rented through private landlords or agents, with 1,452 properties classed as private rented or other and 1,229 properties known to be owned outright without mortgage and considered likely to be rented.[18]

This gave a predicted licensable figure of 35,799 properties. By the end of 2016, Newham had issued over 38,000 licenses,[4] exceeding their target which was not considered to be an absolute figure because the housing market is constantly in flux, but something by which to measure one's success.

It is too early to determine the extent to which public health has been protected through compliance; the collaborative strategy is preventative, enabling identification of the worst properties. The intermediate effect of the strategy is that noncompliant landlords are subjected to heavier regulation than compliant landlords. Over the life of the first five-year licensing scheme ASB has reduced across the borough but remains significantly higher in HMOs and the private rented sector than other tenures.[4]

Deprivation in Newham remains high with overcrowding in 30% of households,[4] and nationally one of the highest rates of homelessness amongst households increasing from 97 per year in 2011 to 1,320 in 2015.[4,19,20] Between the same years, life expectancy in men and women has seen a slight increase but continues to be lower than the England average. Equally, the number of children living in poverty has decreased from 32,370 in 2011 to 19,705 but is still unacceptably significantly worse than the England average.[4,19,20]

Licensing is seen as a key part of the borough's strategy to tackle deprivation, which is viewed as being driven partly by housing and criminality. Licensing allows Newham to target properties with significant ASB and take enforcement action as a first course of action where breaches of the law are witnessed, involving ASB and significant risks to public health such as overcrowding and poor housing conditions.[4]

Sections 60 and 84 of the act make clear that the operation of licensing schemes must be reviewed by local authorities from 'time to time.'[5] I submit that a policy analysis at the midpoint of a five-year designation is a useful method of assessment.

Table 5.2 Axes of Implementation Analysed Using the Dimensions of Cost, Feasibility and Equitability

Axis and Dimensions		Inputs	Outputs	Outcomes
Implementation	Cost	Net cost £329,000 2012/2013 Investment in new staff/training Legal fees 4,219 HMOs and 6,000 other dwellings, monitored for council tax arrears Overall investment to tackle ASB and enviro-crime	Estimated approximate revenue 2012 £3.9 million £17 million (annual expenditure)	Staff costs supported via income Legal fees recovered following successful prosecution Additional council tax over £5million (collected from HMO landlords) Potential saving to NHS in reduction of treatment for housing-related illness unknown
	Feasibility	Police officers already based at Newham Established working policies between council and Metropolitan Police Housing Act 2004 legislation	Established working protocols implemented Legislation used to designate licensing areas	Partnership working Possible Borough-wide licensing implemented
	Acceptability	Consultation carried out with landlord's tenants and private residents Policymakers concerned over transience in borough and tenants being less socially invested Large amount of waste generated in private rented sector	Online licensing system introduced Licence conditions includes: requirements for tenancy agreements Landlords to seek references Restriction on occupancy levels Duties to ensure waste is suitably contained	Security of tenure More social investment from tenants Less churn in private rented sector Reduction of waste in gardens and fly tipping – or prosecution for breach

References

1. Oatt P. *What are the ways in which collaborative working in licensing of private rented housing in the London Borough of Newham could be most effectively used to tackle the worst effects of poor housing?* Dissertation (MSc) – London School of Hygiene and Tropical Medicine, 2017.
2. Morestin F. A framework for analyzing public policies: practical guide. National collaborating centre for healthy public policy [online]. (September 2012). Available from: www.ncchpp.ca/docs/Guide_framework_analyzing_policies_En.pdf [Accessed 20th April 2016.]
3. Dick I. London Borough of Newham Cabinet Report, Private rented property licensing [online]. (June 2012). Available from: www.landlords.org.uk/sites/default/files/$cabinetreportfinal.doca_.ps_.pdf [Accessed 8th September 2019.]
4. London Borough of Newham. Rented property licensing, proposal report for consultation [online]. (October 2016). Available from: www.newham.gov.uk/Documents/Housing/RentedPropertyLicensingProposalConsultation.pdf [Accessed 26th August 2019.]
5. legislation.gov.uk. *Housing Act 2004* [online]. Available from: www.legislation.gov.uk/ukpga/2004/34/contents [Accessed 8th September 2019.]
6. legislation.gov.uk. *Housing Act 1985* [online]. Available from: www.legislation.gov.uk/ukpga/1985/68/pdfs/ukpga_19850068_en.pdf [Accessed 21st December 2016.]
7. London Borough of Newham. Private rented property licensing guide for landlords and managing agents [online]. (June 2012). Available from: www.Newham.gov.uk/Documents/Housing/PropertyLicensingGuideLandlordsAndManagingAgents.pdf [Accessed 21st December 2016.]
8. Tacagni R. Tackling rogue landlords: analysis of private rented housing prosecutions in London. London Property Licensing [online]. (12th May 2015). Available from: www.londonpropertylicensing.co.uk/tackling-rogue-landlords-analysis-private-rented-housing-prosecutions-london [Accessed 4th January 2016.]
9. London Borough of Newham. Borough wide landlord licensing (BWLL) EQIA Template [online]. (1st June 2012). Available from: www.Newham.gov.uk/Documents/Council%20and%20Democracy/BoroughWide-LandlordLicensingEqIA.doc [Accessed 21st December 2016.]
10. London Borough of Newham. Newham info housing reports (Census Data 2011) [online]. (2019). Available from: www.newham.info/housing/report/view/68ccddd9934945b4bd4d518c6fd89263/E01003622/ [Accessed 22nd September 2019.]
11. Renshaw R. *Newham finally wins right to carry on with licensing scheme – after a fight.* Property Industry Eye [online]. (5th December 2017). Available from: https://propertyindustryeye.com/newham-council-wins-the-right-to-carry-on-with-its-licensing-scheme-after-a-fight/ [Accessed 22nd August 2019.]

12. London Borough of Newham. *Newham 2027 Newham's Local Plan the Core Strategy* [online]. (April 2013). Available from: www.Newham.gov. uk/Documents/Environment%20and%20planning/CoreStrategy2004-13. pdf [Accessed 21st December 2016.]

13. Greater London Authority. Indices of Deprivation 2007, Ward Level Summary. [online]. (2007). Available from: https://files.datapress.com/london/dataset/indices-deprivation-2007-ward-level-summary/gla-deprivation-indices-ward-2007.xls [Accessed 28th December 2016.]

14. London Borough of Newham. Who should pay council tax. London Borough of Newham [online]. (2016). Available from: www.Newham.gov. uk/Pages/Services/Council-Tax-who-should-pay.aspx#WhatisCouncil-Tax [Accessed 28th December 2016.]

15. Mishkin P, Moffatt R. A review of multi-agency enforcement and discretionary property licensing to tackle Newham's private rented sector. In Stewart J (ed.) *Effective Strategies and Interventions: Environmental Health and the Private Housing Sector*. London, England: Chartered Institute of Environmental Health, 2013, pp. 12–14.

16. London Borough of Newham. *26,000 applications made by landlords for Newham's property licence scheme* [online]. (31st January 2013). Available from: www.newham.gov.uk/Pages/News/26000-applications-made-by-landlords-for-Newhams-property-licence-scheme.aspx [Accessed 8th September 2019.]

17. Quinn A. *London Borough of Newham: HMO landlord council tax counter fraud project* [online]. (6th June 2019). Available from: www.local.gov.uk/our-support/efficiency-and-income-generation/counter-fraud-hub-case-studies/london-borough-newham [Accessed 8th September 2019.]

18. King J. *Newham Council recoups £5m council tax from rogue landlords*. Newham Recorder [online]. (17th October 2018). Available from: www.newhamrecorder.co.uk/news/newham-council-recoups-5million-council-tax-from-landlords-licensing-scheme-1-5739749 [Accessed 22nd August 2019.]

19. London Borough of Newham. Newham Statistics, data views, census 2011, tenure [online]. (2011). Available from: https://fingertips.phe.org.uk/profile/health-profiles/area-search-results/E12000007?search_type=list-child-areas&place_name=London [Accessed 14th September 2019.]

20. Public Health England. Newham unitary authority health profile 2015 [online]. (2nd June 2015). Available from: https://fingertips.phe.org.uk/profile/health-profiles/area-search-results/E12000007?search_type=list-child-areas&place_name=London [Accessed 14th September 2019.]

21. Mayhew L, Harper G, Waples S. The London Borough of Newham Population growth and change 2007 to 2011 [online]. (August 2011). Available from: www.newham.gov.uk/Documents/Misc/Research-Population2011. pdf [Accessed 21st December 2016.]

22. London Borough of Newham. Waste service information for private landlords [online]. (2016). Available from: www.newham.gov.uk/Pages/ServiceChild/Waste-service-information-for-private-landlords.aspx [Accessed 21st December 2016.]

23. National Landlords Association. Response to the London Borough of Newham Council's Borough-wide Property Licensing: Proposal report for consultation [online]. (April 2012). Available from: www.landlords.org.uk/sites/default/files/Newham%20Selective%20Licensing%20Main%20Consultation%20Response_0.pdf [Accessed 26th December 2016.]

24. Residential Landlords Association. Re: Selective licensing and additional houses in multiple occupation (HMO) consultation [online]. (April 2012). Available from: www.rla.org.uk/docs/rla_newham_consultation_response_0412.pdf [Accessed 29th December 2016.]

6   A small literature review

A literature search of peer-reviewed and grey literature was carried out using a search strategy where the targeted health problem was defined as poor health arising from living in poor housing conditions. The population of concern was tenants in the private rented sector. The intervention was defined as partnership, multi-agency working and property licensing. The outcome was defined as reduction of health risks to occupiers through enforcement and compliance. Searches of titles and abstracts were further refined beginning from 2000 onwards in the Cochrane, IBSS, NICE, PubMed, Social Policy and Practice and Web of Science databases.[1]

The literature on collaborative working in licensing enforcement is sparse because the associated legislation of the Housing Act 2004 is relatively new and its licensing designations have not been used widely enough to generate sufficient study. To examine collaborative working methods across multi-agencies applied to housing, it is necessary to examine other public health initiatives brought about through government-led local area agreements (LAAs) driven by the need to address inequalities across wider remits that encapsulate local strategic partnerships (LSPs) working in housing interventions that are often linked to other social determinants, because there is a greater evidence base for these studies.[1]

The Marmot review[2] argues national policies such as LAAs will fail without effective local delivery systems that strengthen partnerships key to making intervention strategies work across a large scale. Both Marmot[2] and Iles[4] agree that engagement of senior personnel is crucial to service delivery through strong working partnerships where aims and objectives are established at the outset of collaboration. Clarity over the roles of different agencies involved is essential, establishing necessary behaviours for partners to adopt in the delivery of outcomes.

Sandoul and Pipe carried out a recent qualitative survey amongst 20 local authorities who have designated licensing schemes.[5] Several local authorities found aspects of the designation process to be *'disproportionately bureaucratic and costly,'* potentially discouraging some local authorities from adopting selective licensing. Some local authorities reviewed their licensing fees on renewal because the original fee was not reflective of cost.[5]

Newham invested heavily in enforcement across services to tackle housing problems offsetting the set-up costs of the scheme against the first two years.[6] But it had drastically underestimated the size of their private rented sector.[7] However, Newham didn't start with a borough-wide scheme, rather a small pilot scheme in Little Ilford in 2010, in partnership with the police, primary care and the local community, resulting in over 30 prosecutions. In this way, they were able to develop strong working partnerships and apply their learning to the borough-wide scheme.[8,9]

Sandoul and Pipe find agreement with Marmot and Iles, that clarity is needed over outcomes and intentions that must be measured and monitored, suggesting use of non-licensed areas as a baseline measure.[2,4,5] Their findings are also consistent with the policy analysis, agreeing that tangible effects from licensing may not be apparent for several years.[2,4,5]

Selective licensing schemes used as part of a collaborative strategy to address inequalities is discussed in several papers,[10–12] where it is argued that licensing when incorporated into broader public health strategy positively impacts on mental health through enforcement of better housing conditions.[11]

Two studies suggest this claim should be treated with caution, because whilst existing studies point towards physical and mental health gains from housing improvements, the findings are inconsistent and not always statistically significant owing to small study samples, short follow-up times, lack of statistical power, as well as interventions that are categorised too broadly with lack of randomisation or blinding and skewed results due to control groups receiving the intervention. Whilst housing enforcement and compulsory licensing conditions safeguard tenants against physical injury, they do not directly address links between poor housing conditions and mental health.[10,13]

These studies agree on the need for future evaluation of the effectiveness of housing and landlord licensing schemes in addressing health inequalities linked to poor housing,[10,12,13] as many previous housing health-related studies were qualitative in nature[14] and inconclusive due to weak study designs. It is argued that more empirical research is required to determine how collaborative methods can address the links between poor housing and mental health.[10,12,13]

The Marmot report and Joint Strategic Needs Assessments (JSNA) made by Newham and across the South East all concur that socioeconomic status is recognised as an important predictor of population health, where income level and safe comfortable housing quality are considered among wider determinants of health across the social and physical environment.[2,15,16]

The Index of Multiple Deprivation (IMD) for 2015 measures deprivation at ward level by proportion of lower super output areas (LSOA) that lie within each decile scoring between 1 and 10; 1 represents the most deprived 10% of LSOAs in England and 10 the least deprived 10%. The majority of LSOAs in Newham scored between 1 and 3 making up 80% of the borough. As much as 73% of LSOAs scored between 1 and 3 for income deprivation.[17] Income is considered to be a socioeconomic resource that individuals can use to upgrade their living conditions by paying for better ones. But low-income earners like a family living in a shed cannot always afford to escape deprivation in this way.[6,16–18]

It is argued by Marmot and others that public health collaborations inclusive of housing policy should be based around the social determinants of health, identifying collaborative advantages.[2,19,20] Such policy should be evidence based and translated into future working practices to address inequalities. For this purpose, policymakers are encouraged to make efficient use of Joint Strategic Needs Assessments (JSNAs) to develop strategies with integrated services incorporating housing policy, social care, public health and the NHS.[2,20]

Edwards concurs that local authorities could help reduce inequalities through improved collaboration and shared practice, examining the extent to which JSNAs have been applied to housing interventions, concluding that few cases showed any evidence basis of cross-referencing to wider determinants such as the socioeconomic factors of deprivation, ethnicity, age or geographical location.[16] Edwards finds agreement with Newham's core plan in stressing that data should be measurable at ward level.[16,18]

Sandoul and Pipe state that some local authorities find the commissioning of stock condition surveys as part of their justification for licensing to be costly,[5] but Edwards sees these surveys as contributory to JSNAs and a local authorities overall housing strategy on health and well-being issues such as thermal efficiency, safety, size and costs of repairs.[16] In developing housing strategy, Edwards states local authorities should make use of available local-level statistics and strategic housing marketing assessments.[16] This finds agreement with emphasis within sections 57 and 81 of the act that exercising of licensing powers must be consistent with the local authorities overall housing strategy in combination with other available courses of action.[3]

Newham's Equality Impact Assessment (EQIA) of June 2012[6] showed their intelligence on the private rented sector was 'imperfect' prior to licensing. Measuring deprivation or overcrowding using existing data from conventional sources such as health surveys, census or council tax or electoral registers data will initially yield conservative estimates of the true figure. But licence application data with property details are measurable at ward level.[7]

Finney et al. and a second study by Baker Mitchell and Pell measured ethnicity, occupation, education, housing tenure and income; the first was a quantitative study of census and IMD data. The second was a cross-sectional study using individual and area-wide measures to target limited public health resources efficiently guided by the 2004 IMD. Both studies identified areas with higher levels of deprivation amongst ethnic minority groups, across the country. The latter study concluded that ethnic groups would not be disadvantaged by efficient area-based targeting.[21,22]

These studies all demonstrate that housing inequalities are a nationwide concern, illustrating that wider measures should be adopted to distinguish deprivation levels when designing public health interventions aimed at reducing socioeconomic inequalities. Overall, the themes from these studies discuss policy and area-wide strategies, statistics and data usage for policy implementation, and the need for community engagement. The policy analysis showed that it is necessary to draw on all these aspects to develop sustainable outputs that result in lasting physical improvements. There needs to be clarity on what you want to achieve and the action you need to take.

Sandoul and Pipe found that informal action is still heavily relied upon by local authorities with licensing designations who view enforcement as a *'backstop'* to tackle noncompliance instead of being used as a first course of action.[5] Some local authorities are using the threat of removing or reducing the licence if landlords do not carry out repairs, but in Newham's case this threat is backed up by a strong enforcement approach to serious hazards. Other local authorities appear to be using numerous warning letters (lines in the sand) and other informal resolutions instead of applying the law when they find unlicensed properties. Several local authorities with licensing evidence a combined approach through service of notices on identification of category 1 and high category 2 hazards.

Devising a strategic housing intervention is messy; whether it's a designation or just a bed in a shed, if you don't think it through properly you are in danger of repeating the mistakes of the past. The intervention required to resolve the problem uncovered in Forest Gate involved the collaboration of the police, immigration, council tax, benefits and housing options to bring about a satisfactory resolution. It all took place amidst a harsh financial climate of ring fencing and budget reductions, all of which makes it difficult to sustain partnerships when other agencies are being cut.[19,23–25] Newham were also going through a period of uncertainty,[26] whilst waiting for the Secretary of State to give permission on renewal of its five-year additional licensing scheme. My job, and many of my colleagues', depended on this, and at the same time, our law enforcement partners had been under siege through cuts since 2010.[27]

Iles and Geddes[4,23] agree that in order to sustain partnerships under external pressures they should be treated as common threats beyond the control of the partnership but that contending parties should be required to respond using collaborative strategies. It is failure to acknowledge this reality that subsequently undermines unity and working attitudes.

Barriers to collaboration do not just arise because of the threat that your service is going to get cut. Public health services have lived with this threat almost daily since at least the 1980s when they underwent service reformations that introduced a purchaser–provider split with target-driven competition amongst agencies that were fragmented to develop more fluid structures and relationships. This competition amongst agencies made it difficult to develop collaboration or establish a common identity during subsequent introductions of area-based initiatives such as LSPs. The initiatives were reliant upon pooled funding or competition for funding to deliver services over wide remits amongst weak, poorly managed partnerships lacking in resources, decision-making capacity and communicative ability.[14,23,28]

Stewart and Battersby agree that diminishing resources available to local authorities and Environmental Health Departments is one of the reasons that local authorities are not maximising the use of their enforcement powers under the act.[29,30] But that there are opportunities to work together with public health to use evidence-based practice and research to promote health initiatives. Agreement is found in Buck and Dunn,[31] who observed that there is a lack of published evidence from local authorities on the cost-effectiveness of environmental interventions, because they are statutory duties. The study argues that the need for published evidence is now greater because local authorities are experiencing financial cuts leading to difficult decisions around prioritisation of services.

Sandoul and Pipe argue that Civil Financial Penalty Notices (FPNs) are an opportunity to balance budgets against the running costs of selective licensing.[5] But this has to be cost-effective, and the financial penalty has to be justifiably punitive, backed up through investigation into the offender's finances.[32]

The case study demonstrates that it is not always possible to obtain accurate financial information. In taking the decision to impose the penalties totalling £15,000, we also considered the cost of appeal, officer time in preparing evidence, barrister fees for representation and time spent at tribunal. The £15,000 figure was arrived at largely because of the licence holder's large portfolio. Sandoul and Pipe found many local authorities have learned more about their private rented sectors through licensing than was previously known; this was also Newham's finding.[5,7]

The licence holder's Newham properties were known to us, and it was easy to assess his portfolio and arrive at a justifiable sum. The agency was relatively unknown, and there was no real way to assess their portfolio. It emerged in the tribunal that I had underestimated their fine because they manage 50 properties.[33] The overall fine was reduced to £10,000 because the licence holder's reasonable excuse was accepted, but this was still a decent financial return for our efforts.

If the fines are high then it is worth applying a civil FPN subject to a cost-benefit analysis; if the fine is too low and the cost of a tribunal outweighs it, then consideration should be given to prosecution instead, and the possibility of an unlimited fine, or service of a notice which might mean returning to do a Health and Housing Safety Rating System (HHSRS) inspection.

The documentation used to carry out the policy analysis, and the analysis itself, finds agreement with these studies that collaborative intervention activity needs strong coordination, dedicated resources and a framework for multi-agency operation. The key to bringing all this together when using licensing powers is to be able to identify the inequalities and social determinants of health at local level and translate them into policy before developing strong collaborative partnership working,[2,6,34] but this alone is insufficient without marrying it all to the prevalent levels of antisocial behaviour (ASB) and housing criminality that satisfies sections 57 and 81 of the act.[3] But the best way is to start licensing enforcement with a small pilot scheme and refine the working practices.

The lesson learned at Newham was that to move policy on housing and health forward, you have to look at moving it first on housing and criminality. The case study on the shed exemplifies this because in order to reduce the risk to the occupier's health it was necessary to take enforcement action over the shed, the broken window and licensing. In so doing, we have better health outcomes as a derived benefit, for Maya, her husband and child, as well as the subletting family.

Other types of housing interventions such as LAAs and LSPs are directed under top-down central government measures. A 2004 report to the House of Commons found these interventions had poor management, weak partnerships and lack of community engagement.[35] Selective licensing works because it is developed at local level.[5–7,34] This finds agreement with Marmot,[7] Edwards[16] and Baker Mitchell and Pell.[22]

Allen shows us that poor understanding of aims and objectives and roles in partnership impedes collaborative working.[14] To reach the point where there is collaborative advantage with mutual aims and outcomes as argued by Iles and Marmot, one has to have prepared a strategic evidence-based policy upon which partnerships can develop.[2,4]

When collaborating, partners prioritise their own goals; instead, they should be open to joint goals that may not be apparent at first but emerge through close working, making the partnership co-evolutionary, holistic and more collaborative across multi-agencies.[4,14,25] The continued gathering and sharing of intelligence and data in preparation ahead of Newham's collaborative operations and the carrying out of these operations have sustained the partnerships.

Selective licensing schemes have been criticised for partnering with Immigration and Customs Enforcement. Through conservative census figures, the policy analysis shows that police and immigration arrests in Newham occurred amongst only 5.56% of the households inspected at the midpoint of their first five-year scheme. Migration in Newham has always been high, and beneficial to services and the economy.[7]

Poppleton et al. acknowledge the economic contribution that low-skilled migrant workers make but raises concerns about the higher impacts of health through this group being most likely to be living in poor conditions.[36] Irrespective of whether or not such migration is legal, gross inequalities in health through poor housing is unacceptable. It should be addressed through adequate health and social measures that demand coordinated efforts allowing access to public health services for all, in the spirit of the Alma Ata declaration of 1978.[37]

Criminal landlords find it advantageous to house people who have little choice over their living conditions, and illegal eviction or tenant harassment is seen as an unintended effect of licensing regulation.[6] But unlike the House of Commons findings in 2004,[35] that highlighted poor working practices amongst partnerships, this phenomenon is occurring through the poor working practices of noncompliant landlords, whose managerial behaviour licensing schemes are intended to change.[6,7,34]

The new Deregulation Act 2015 provides extra protection from section 21 evictions where local authorities have inspected under HHSRS and issued a 'relevant notice' which is defined under section 13 as either an improvement or an emergency remedial notice.[38] Sandoul and Pipe were unable to draw conclusions on its effectiveness and what relationship it bears to reliance on informal action.[5] But section 21 applies only to assured short-hold tenants,[39] and this power is of no use to undocumented migrants at the mercy of criminal landlords.

The answer to this issue lies in partners recognising the problems and sustaining high levels of collaborative enforcement as a deterrent that will eventually bring about a change in culture among this type of landlord.[6]

Enforcement decisions need to be based upon cost analysis taking into account costs for appeals in courts or tribunals and combining licensing enforcement with follow-up HHSRS and part 1 enforcement options to address any remaining hazards.

Collaboration in private rented sector licensing between local authorities and agency partnerships is sustained through working towards collaborative advantages. This ensures that the licensing fees, costs of running the scheme and enforcement outweigh any risks. Local authorities and public health should work together to explore new areas of collaboration where licensing is incorporated into local health strategies, with good use of data as part of a local housing strategy.

A wider evidence base needs to be developed using empirical research to demonstrate the effectiveness of licensing schemes to policymakers to assist with the future designation of such schemes across other areas. Such study designs should be carried out across local authorities with five-year schemes, making use of available public health data and in comparison with unlicensed boroughs with comparable IMDs with an agreed consistent method of data collection at ecological and cross-sectional levels collecting information on multiple, aetiologically relevant exposures and their outcomes rather than broad categorisations, with sufficient study numbers followed up for the five-year duration across boroughs, ensuring temporality of association and results that are statistically significant.

References

1. Oatt P. *What are the ways in which collaborative working in licensing of private rented housing in the London Borough of Newham could be most effectively used to tackle the worst effects of poor housing?* Dissertation (MSc) – London School of Hygiene and Tropical Medicine, 2017.
2. Marmot M. *Fair society, healthy lives: the Marmot Review: strategic review of health inequalities in England post 2010*, 2010.
3. legislation.gov.uk. *Housing Act 2004* [online]. Available from: www. legislation.gov.uk/ukpga/2004/34/contents [Accessed 27th February 2016.]
4. Iles V, Vaughan Smith J. *The learning from the beyond partnership learning set 2004–2006* [online]. (March 2006). Available at: www.reallylearning. com/BeyondPartnership.pdf [Accessed 18th September 2016.]
5. Sandoul T, Pipe D. *A Licence to Rent*. Chartered Institute of Housing and Chartered Institute of Environmental Health [online]. (January 2019). Available from: www.cih.org/resources/PDF/Policy%20free%20 download%20pdfs/A%20Licence%20to%20Rent%20-%20selective%20 licensing.pdf [Accessed 21st September 2019.]
6. London Borough of Newham. Borough wide landlord licensing (BWLL) EQIA Template. [online]. (1st June 2012). Available from: www.Newham.gov. uk/Documents/Council%20and%20Democracy/BoroughWideLandlordLi-censingEqIA.doc [Accessed 21st December 2016.]
7. London Borough of Newham. Rented property licensing, proposal report for consultation. [online]. (October 2016). Available from: www.newham. gov.uk/Documents/Housing/RentedPropertyLicensingProposalConsul-tation.pdf [Accessed 26th August 2019.]
8. Hammond-Laing R. *Selective licensing comes to London!* Lacors [online]. (21st December 2009). Available from: www.ihsti.com/lacors/ContentDetails. aspx?id=22946 [Accessed 26th August 2019.]
9. London Borough of Newham. Newham private rented property licensing scheme: the enforcers go in [online]. (28th February 2013). Available from: www.Newham.gov.uk/Pages/News/Newham-private-rented-property-licensing-scheme-the-enforcers-go-in.aspx [Accessed 1st August 2018.]
10. Barrat C, Kitcher C, Stewart J. Beyond safety to wellbeing: how local authorities can mitigate the mental health risks of living in houses in multiple occupation. *Journal of Environmental Health Research* 2012; 12(1): 39–49.
11. Clayton C. A problem shared. *Health Service Journal* February 2012; 19(1): 1227–1227.
12. Public Health England. Due North: The report of the inquiry on health equity for the North. University of Liverpool and Centre for Local Economic Strategies, 2014.
13. Curl A, Kearns A, Mason L, et al. Physical and mental health outcomes following housing improvements: evidence from the GoWell study. *J Epidemiol Community Health* 2015; 69: 12–19.
14. Allen T, Improving housing, improving health the need for collaborative working. *British Journal of Community Nursing* 2006; 11(4): 157–161.

15. London Borough of Newham and NHS Newham. Newham joint strategic needs assessment 2011/12, September 2012 Update [online]. (September 2012). Available from: www.Newham.gov.uk/Documents/Council%20and%20Democracy/JSNASept2012Update.pdf [Accessed 21st December 2016.]

16. Edwards M. *Joint strategic needs assessment and housing: report of a study on the South East region.* Housing, Learning & Improvement Network, 2009.

17. Greater London Authority. London datastore. London area profiles – Newham [online]. (2019). Available from: https://iao.blob.core.windows.net/publications/reports/f11c199d237c4cb79bca5427bfe8511d/E09000025.html [accessed 22nd September 2019.]

18. London Borough of Newham. *Newham 2027 Newham's local plan the core strategy* [online]. (April 2013). Available from: www.Newham.gov.uk/Documents/Environment%20and%20planning/CoreStrategy2004-13.pdf [Accessed 21st December 2016.]

19. Riches N, Coleman A, Gadsby E et al. *The role of local authorities in health issues: a policy document analysis.* Policy Research Unit in Commissioning and the Healthcare System, 2015.

20. Oxford Brookes University. *Institute of public care. Health, wellbeing, and the older people housing agenda: briefing paper.* Housing, Learning & Improvement Network, 2012.

21. Finney N, Lymperopolou K, Kapoor N, et al. *Local ethnic inequalities: ethnic differences in education, employment, health and housing in districts of England and Wales, 2001–2011.* Social Policy and Practice February 2016; 57.

22. Baker J, Mitchell R, Pell J. Cross-sectional study of ethnic differences in the utility of area deprivation measures to target socioeconomically deprived individuals. *Social Science & Medicine* 2013; 85: pp. 27–31.

23. Geddes M, Davies J, Fuller C. Evaluating local strategic partnerships: theory and practice of change. *Local Government Studies* February 2007; 33(1): 97–116.

24. Beatty C, Foden C, Lawless P et al. Area-based regeneration partnerships and the role of central government: the New Deal for Communities programme in England. *Policy and Politics* 2010; 38(2): 235–251.

25. Riches N, Coleman A, Gadsby E et al. *The role of local authorities in health issues: a policy document analysis.* Policy Research Unit in Commissioning and the Healthcare System, 2015.

26. Wales R, Healey J. *Sajid Javid should give councils free rein to tackle rogue landlords.* The Guardian [online]. (19th September 2017). Available from: www.theguardian.com/housing-network/2017/sep/19/sajid-javid-councils-tackle-rogue-landlords-newham [Accessed 23rd August 2019.]

27. Travis A. *Simple numbers tell story of police cuts under Theresa May.* The Guardian [online]. (5th June 2017). Available from: www.theguardian.com/uk-news/2017/jun/05/theresa-may-police-cuts-margaret-thatcher-budgets [Accessed 23rd August 2019.]

28. Matthews P. (2014) Being strategic in partnership - interpreting local knowledge of modern local government. *Local Government Studies* May 2014; 40(3): 451–472.

29. Stewart J. The environmental health practitioner: new evidence-based roles in housing, public health and well-being. *Perspectives in Public Health* November 2013; 133(6): 325–329.

30. Battersby S. *The challenge of tackling unsafe and unhealthy housing* [online]. (December 2015) Available from: http://sabattersby.co.uk/documents/KBReport2.pdf [Accessed 27th February 2016.]

31. Buck D, Dunn P. (2015) *The district council contribution to public health: a time of challenge and opportunity*. The Kings Fund 2015.

32. Ministry of Housing Communities and Local Government. Civil penalties under the Housing and Planning Act 2016. Crown Copyright. [online]. (2018) Available from: https://assets.publishing.service.gov.uk/government/uploads/system/uploads/attachment_data/file/697644/Civil_penalty_guidance.pdf [Accessed 1st August 2018.]

33. LON/00BB/HNA/2017/0018 & 19. Residential property tribunal decisions. First-Tier tribunal property chamber (residential property) [online]. (1st March 2019). Available from: https://assets.publishing.service.gov.uk/media/5c7916d2ed915d29eb6a0056/LON00BBHNA20170018___19_-_Bristol_Road.pdf?_ga=2.251520774.1464556537.1566800783-304649890.1566047204 [Accessed 23rd August 2019.]

34. Dick I. London Borough of Newham Cabinet Report, private rented property licensing. (June 2012) [online]. Available from: www.landlords.org.uk/sites/default/files/$cabinetreportfinal.doca_.ps_.pdf [Accessed 21st December 2016.]

35. House of Commons Committee of Public Accounts. *An early progress report on the new deal for communities programme*. The House of Commons, London, 2004.

36. Poppleton S, Hitchcock K, Lymperopoulou K, et al. *Social and public service impacts of international migration at the local level*. Assets.publishing.service.gov.uk. [online]. (2013) Available from: https://assets.publishing.service.gov.uk/government/uploads/system/uploads/attachment_data/file/210324/horr72.pdf [Accessed 23rd August 2019.]

37. Declaration of Alma-Ata. International Conference on Primary Health Care, Alma- Ata, USSR, 6–12 September 1978. [online]. Available from: www.who.int/publications/almaata_declaration_en.pdf?ua=1 [Accessed 23rd September 2019.]

38. legislation.gov.uk. *Deregulation Act 2015* [online]. Available from: www.legislation.gov.uk/ukpga/2015/20/contents/enacted[Accessed23rdSeptember 2019.]

39. Gov.UK. *Evicting tenants (England and Wales)* [online]. Available from: www.gov.uk/evicting-tenants/section-21-and-section-8-notices [Accessed 24th September 2019.]

Index

Note: **Bold** page numbers refer to tables and *italic* page numbers refer to figures.